Graduate Texts in Mathematics

69

Serge Lang

Cyclotomic Fields II

Springer-Verlag

New York Heidelberg Berlin

Serge Lang
Department of Mathematics
Yale University
New Haven, Connecticut 06520
USA

Editorial Board

P. R. Halmos

Managing Editor
Department of Mathematics
Indiana University
Bloomington, Indiana 47401
USA

F. W. Gehring

Department of Mathematics
University of Michigan
Ann Arbor, Michigan 48104
USA

C. C. Moore

Department of Mathematics
University of California
Berkeley, CA 94720
USA

AMS Subject Classification (1980): 12A35

Library of Congress Cataloging in Publication Data

Lang, Serge, 1927–
  Cyclotomic fields II.

  (Graduate texts in mathematics; v. 69)
  Bibliography: p.
  Includes index.
  1. Fields, Algebraic, 2. Cyclotomy. I. Title.
II. Series.
QA247.L34   512′.3   79-20459

ISBN 0-387-90447-6 Springer-Verlag New York
ISBN 3-540-90447-6 Springer-Verlag Berlin Heidelberg

Preface

This second volume incorporates a number of results which were discovered and/or systematized since the first volume was being written. Again, I limit myself to the cyclotomic fields proper without introducing modular functions.

As in the first volume, the main concern is with class number formulas, Gauss sums, and the like. We begin with the Ferrero–Washington theorems, proving Iwasawa's conjecture that the p-primary part of the ideal class group in the cyclotomic $\mathbf{Z}_p$-extension of a cyclotomic field grows linearly rather than exponentially. This is first done for the minus part (the minus referring, as usual, to the eigenspace for complex conjugation), and then it follows for the plus part because of results bounding the plus part in terms of the minus part. Kummer had already proved such results (e.g. if $p \nmid h_p^-$ then $p \nmid h_p^+$). These are now formulated in ways applicable to the Iwasawa invariants, following Iwasawa himself.

After that we do what amounts to "Dwork theory," to derive the Gross–Koblitz formula expressing Gauss sums in terms of the p-adic gamma function. This lifts Stickelberger's theorem p-adically. Half of the proof relies on a course of Katz, who had first obtained Gauss sums as limits of certain factorials, and thought of using Washnitzer–Monsky cohomology to prove the Gross–Koblitz formula.

Finally, we apply these latter results to the Ferrero–Greenberg theorem, showing that $L_p'(0, \chi) \neq 0$ under the appropriate conditions. We take this opportunity to introduce a technique of Washington, who defined the p-adic analogues of the Hurwitz partial zeta functions, in a way making it possible to parallel the treatment from the complex case to the p-adic case, but in a much more efficient way.

All of these topics form a natural continuation of those of Volume I. Thus chapters are numbered consecutively, and the bibliography (suitably expanded) is similarly updated.

I am much indebted to Larry Washington and Neal Koblitz for a number of suggestions and corrections; and to Avner Asch for helping with the proofreading.

Larry Washington also read the first volume carefully, and made the following corrections with no other changes in the proofs:

Chapter 5, Theorem 1.2(ii), p. 127: read $e_n = dn + c_0$ for some constant c_0.

Chapter 7, Theorem 1.4, p. 174: the term $1/k^2$ should be $(-1)^k/k \cdot k!$ instead.

Chapter 8, Formulas **LS 6**, p. 207: one needs to assume that $[\pi](X)$ is a polynomial. This is satisfied if the formal group is the basic Lubin–Tate group, and the theorems proved are invariant under an isomorphism of such groups, so the proofs are valid without further change.

Washington also pointed out the reference to Vandiver [Va 2], where indeed Vandiver makes the conjecture:

> ... However, about twenty-five years ago I conjectured that this number was never divisible by l [referring to h^+]. Later on, when I discovered how closely the question was related to Fermat's Last Theorem, I began to have my doubts, recalling how often conjectures concerning the theorem turned out to be incorrect. When I visited Furtwängler in Vienna in 1928, he mentioned that he had conjectured the same thing before I had brought up any such topic with him. As he had probably more experience with algebraic numbers than any mathematician of his generation, I felt a little more confident ...

On the other hand, many years ago, Feit was unable to understand a step in Vandiver's "proof" that $p \nmid h^+$ implies the first case of Fermat's Last Theorem, and stimulated by this, Iwasawa found a precise gap which is such that the proof is still incomplete.

New Haven, Connecticut　　　　　　　　　　　　　　　　SERGE LANG
1980

Contents Volume II

Contents Volume I

Notation

As in the first volume, if A is an abelian group and N a positive integer, we let A_N be the kernel of multiplication by N, and

$$A(N) = A/NA.$$

If p is a prime, we let $A^{(p)}$ be the subgroup of p-primary elements, that is, those elements annihilated by a power of p.

Measures and Iwasawa Power Series

10

This chapter gives a number of complements to Chapter 4. In §1 we extend the formalism of the associated power series to the change of variables

$$x \leftrightarrow \gamma^x$$

for $x \in \mathbf{Z}_p$ and γ equal to a topological generator of $1 + p\mathbf{Z}_p$. A measure on $1 + p\mathbf{Z}_p$ then corresponds to a measure on $\mathbf{Z}_p$, and we give relations between their associated power series. This is then applied to express Bernoulli numbers $B_{k,\chi}$ as values of power series. We write

$$\chi = \theta\omega^{-k}\psi = \theta_k\psi,$$

where first θ is an even character on $\mathbf{Z}(dp)^*$ (d prime to p), ω is the Teichmuller character, and ψ is a character on $1 + p\mathbf{Z}_p$. Let $\zeta = \psi(\gamma)$. Then

$$\frac{1}{k} B_{k,\chi} = f_{\theta,k}(\zeta - 1),$$

where $f_{\theta,k}$ depends only on θ and k. This allows a partial asymptotic determination of $\mathrm{ord}_p B_{k,\chi}$ when θ is fixed, and the conductor of ψ tends to infinity, due to Iwasawa [Iw 14], §7. This gives rise to the corresponding asymptotic estimate for the minus part of class numbers of cyclotomic extensions.

The Iwasawa expressions for the Bernoulli numbers gives an asymptotic value for their orders:

$$\mathrm{ord}_p B_{k,\theta_k\psi} = mp^n + \lambda n + c$$

1

for n sufficiently large, cond $\psi = p^{n+1}$. In order that $m \neq 0$, Iwasawa showed that a system of congruences had to be satisfied (essentially that the coefficients of the appropriate power series are $\equiv 0 \pmod{p}$). We derive these congruences here in each case successively. The next chapter is devoted to the proofs by Ferrero–Washington that these congruences cannot all be satisfied, whence the Iwasawa invariant m is equal to 0.

At the end of their paper, Ferrero–Washington conjecture that the invariant λ_p for the cyclotomic $\mathbf{Z}_p$-extension of $\mathbf{Q}(\mu_p)$ satisfies a bound

$$\lambda_p \ll \frac{\log p}{\log \log p}.$$

I am much indebted to Washington for communicating to me the exposition of the steps which lead to this conjecture, and which were omitted from their paper.

§1. Iwasawa Invariants for Measures

We let p be an odd prime for simplicity. The multiplicative group $1 + p\mathbf{Z}_p$ is then topologically cyclic, and we let γ denote a fixed topological generator. Then $\gamma \bmod p^n$ generates the finite cyclic group $1 + p\mathbf{Z}_p \bmod p^n$ for each positive integer n. For instance, we may take

$$\gamma = 1 + p.$$

[*Note*: If $p = 2$, then one has to consider $1 + 4\mathbf{Z}_2$ instead of $1 + 2\mathbf{Z}_2$.]
    There is an isomorphism

$$\mathbf{Z}_p \to 1 + p\mathbf{Z}_p$$

given by

$$x \mapsto \gamma^x.$$

Its inverse is denoted by α, so that by definition

$$\alpha(\gamma^x) = x.$$

Let $d \geq 1$ be a positive integer prime to p. We shall consider measures on the projective system of groups

$$Z_n = \mathbf{Z}(dp^n) = \mathbf{Z}/dp^n\mathbf{Z} = \mathbf{Z}(d) \times \mathbf{Z}(p^n).$$

The projective limit is simply denoted by

$$Z = \mathbf{Z}(d) \times \mathbf{Z}_p.$$

A measure is then determined by a family of functions μ_n on Z_n, as in Chapter 2, §2. We let

$$Z^* = \mathbf{Z}(d) \times \mathbf{Z}_p^* \quad \text{and} \quad Z^{**} = \mathbf{Z}(d)^* \times \mathbf{Z}_p^*.$$

An element $z \in Z^*$ can be written uniquely in the form

$$z = (z_0, \eta\gamma^x) = (z_0, z_p) \quad \text{with } z_0 \in \mathbf{Z}(d), \eta \in \boldsymbol{\mu}_{p-1}, x \in \mathbf{Z}_p.$$

We define the homomorphism

$$\alpha: Z^* \to \mathbf{Z}_p \quad \text{by} \quad \alpha(z_0, \eta\gamma^x) = x.$$

We define as usual

$$\langle z \rangle_p = \langle z \rangle = \langle z_p \rangle = \gamma^x,$$

so that $\alpha(z) = \alpha(\langle z \rangle)$. As above, we usually omit the index p on $\langle z \rangle_p$.

A continuous function on $\mathbf{Z}_p$ gives rise to a continuous function on $1 + p\mathbf{Z}_p$ by composition with α, and conversely.

As in Chapter 2, §1 we let $\mathfrak{o}$ be the ring of p-integers in $\mathbf{C}_p$, and we let μ be an $\mathfrak{o}$-valued distribution, i.e. a measure.

By the basic correspondence between functionals and measures, we obtain the following theorem.

Theorem 1.1. *Let μ be a measure on Z with support in Z^*. Then there exists a unique measure $\alpha_* \mu$ on $\mathbf{Z}_p$ such that for any continuous function φ on $1 + p\mathbf{Z}_p$ we have*

$$\int_{Z^*} \varphi(\langle a \rangle) \, d\mu(a) = \int_{\mathbf{Z}_p} \varphi(\gamma^x) \, d(\alpha_* \mu)(x).$$

We now describe the power series associated with $\alpha_* \mu$ modulo the polynomial

$$h_n(X) = (1 + X)^{p^n} - 1.$$

Thus we fix a value of $n \geq 0$, and for each $a \in Z^*$ we let $r(a)$ be the unique integer such that

$$0 \leq r(a) < p^n \quad \text{and} \quad r(a) \equiv \alpha(a) \bmod p^n.$$

Theorem 1.2. *Let f be the power series associated with $\alpha_* \mu$. Let*

$$Z_{n+1}^* = Z(d) \times Z(p^{n+1})^*$$

Then

$$f(X) \equiv \sum_{a \in Z_{n+1}^*} \mu_{n+1}(a)(1+X)^{r(a)} \bmod h_n(X).$$

Proof. By the definition of the associated power series, we have

$$f(X) \equiv \sum_{r=0}^{p^n-1} (\alpha_* \mu)(r)(1+X)^r.$$

But letting char denote the characteristic function, we have:

$$(\alpha_* \mu)(r \bmod p^n) = \int_{Z_p} (\text{char of } r \bmod p^n) \, d(\alpha_* \mu)$$

$$= \int_{Z^*} (\text{char of } Z(d) \times \mu_{p-1} \times \gamma^{r+p^n Z_p}) \, d\mu$$

(by Theorem 1.1)

$$= \sum_\eta \mu_{n+1}(\eta \gamma^r \bmod p^{n+1})$$

where this last sum is taken over $\eta \in Z(d) \times \mu_{p-1}$. This proves the theorem.

Corollary 1. *Let ψ be a nontrivial character of $1+pZ_p$, with conductor p^{n+1}. Define $\psi(a) = \psi(\langle a \rangle)$. Let*

$$\psi(\gamma) = \zeta = \textit{primitive } p^n\textit{-th root of unity.}$$

Let f be the power series associated with $\alpha_* \mu$. Then

$$\int_{Z_p^*} \psi \, d\mu = f(\zeta - 1).$$

Proof. We have

$$\int_{\mathbf{Z}^*} \psi \, d\mu = \int_{\mathbf{Z}_p} \psi(\gamma^x) \, d(\alpha_* \mu)(x) \qquad \text{(by Theorem 1.1)}$$

$$= \int_{\mathbf{Z}_p} \zeta^x \, d(\alpha_* \mu)(x)$$

$$= f(\zeta - 1). \qquad \text{(by Theorem 1.2 of Chapter 4).}$$

This proves the corollary.

We continue with the same notation as in the theorem. We shall use the notation

$$B(\psi, \mu) = \int_{\mathbf{Z}^*} \psi \, d\mu = f(\zeta_\psi - 1).$$

Suppose that there exists a rational number m such that the power series f can be written in the form

$$f(X) = p^m(c_0 + c_1 X + \cdots + c_{\lambda-1} X^{\lambda-1} + c_\lambda X^\lambda + \cdots)$$

where c_λ is a unit in $\mathfrak{o}$, and $c_0, \ldots, c_{\lambda-1} \in \mathfrak{m}$, the maximal ideal of $\mathfrak{o}$. We call m, λ the **Iwasawa invariants of** μ, or f. If the measure μ has values in the maximal ideal of the integers in a field where the valuation is discrete (which is the case in applications), then f has coefficients in that ring, and such m, λ exist if $f \neq 0$. If $m = 0$, then λ is the Weierstrass degree of f. In any case, λ is the Weierstrass degree of $p^{-m} f$.

As usual, we shall write

$$x \sim y$$

to mean that x, y have the same order at p.

Corollary 2. *There exists a positive integer n_0 (depending only on f) such that if $n \geq n_0$ and cond $\psi = p^n$, then*

$$B(\psi, \mu) \sim p^m(\zeta - 1)^\lambda$$

where ζ is a primitive p^n-th root of unity.

Proof. As $n \to \infty$, the values $|\zeta - 1|$ approach 1, and so the term $c_\lambda(\zeta - 1)^\lambda$ dominates in the power series $f(\zeta - 1)$ above.

Corollary 3. *For some constant* $c = c(f)$, *we have*

$$\operatorname{ord}_p \prod_{\substack{\text{cond } \psi = p^t \\ n_0 \leq t \leq n}} B(\psi, \mu) = mp^n + \lambda n + c(f)$$

Proof. Since

$$\prod_{\substack{\zeta^{p^n} = 1 \\ \zeta \neq 1}} (\zeta - 1) = p^n,$$

the formula is immediate, since the product taken for $n_0 \leq t \leq n$ differs by only a finite number of factors (depending on n_0) from the product taken over all t, and we can apply Corollary 2 to get the desired order.

In the light of Corollary 3, we shall call m the **exponential invariant**, and λ the **linear invariant**.

Let f be as above, the power series associated with $\alpha_* \mu$, and put

$$c_r^{(n)} = \sum_\eta \mu_{n+1}(\eta \gamma^r \bmod p^{n+1}).$$

Then

$$f(X) \equiv \sum_{r=0}^{p^n - 1} c_r^{(n)} (1 + X)^r \bmod h_n$$

$$\equiv \sum_{r=0}^{p^n - 1} a_r^{(n)} X^r \qquad \bmod h_n,$$

where the coefficients $a_r^{(n)}$ are obtained from the change of basis from

$$1, X, \ldots, X^{p^n - 1}$$

to

$$1, 1 + X, \ldots, (1 + X)^{p^n - 1}.$$

We can rewrite $c_r^{(n)}$ in terms of the variable $u = \gamma^r$, namely

$$c^{(n)}(u) = \sum_\eta \mu_{n+1}(\eta u \bmod p^{n+1}).$$

These coefficients $c^{(n)}(u)$ will be called the **Iwasawa coefficients**.

6

Theorem 1.3. *Let n be an integer ≥ 0 such that $c_r^{(n)}$ is a p-unit for some integer r with*

$$0 \leq r \leq p^n - 1.$$

Then the exponential Iwasawa invariant m of μ is equal to 0, and we have $\lambda \leq p^n$.

Proof. Some coefficient $a_r^{(n)}$ must also be a p-unit with r in the same range, and we can write

$$f(X) = \sum_{r=0}^{p^n-1} a_r^{(n)} X^r + g_1(X) X^{p^n} + p g_2(X),$$

where $g_1(X), g_2(X) \in \mathfrak{o}[[X]]$. Hence the coefficient a_r of $f(X)$ is itself a p-unit, whence the theorem follows.

We shall sometimes deal with certain measures derived by the following operation from μ. Let $s \in \mathbf{Z}_p$. We define the s-th **twist** of μ to be the measure defined on Z^* by

$$\mu^{(s)}(a) = \langle a \rangle^s \mu(a),$$

and equal to 0 outside Z^*. In that case, the coefficients $c_r^{(n)}$ should be indexed by s, i.e.

$$c_{r,s}^{(n)} = c_r^{(n)} \gamma^{rs}.$$

Since γ^{rs} is a p-adic unit, it follows that the same power of p divides all $c_{r,s}^{(n)}$ as divides $c_r^{(n)}$. Thus Theorem 1.3 also applies to the twisted measure and the power series f_s associated with $\alpha_*(\mu^{(s)})$ instead of f in the theorem, and we find:

Theorem 1.4. *Let m_s, λ_s be the Iwasawa invariants of $\mu^{(s)}$. If $m_s = 0$ for some s, then $m_s = 0$ for all s. Suppose this is the case, and let n be the positive integer such that*

$$p^{n-1} \leq \lambda_0 < p^n.$$

Then we also have

$$p^{n-1} \leq \lambda_s < p^n$$

for all s.

§2. Application to the Bernoulli Distributions

Let $\mathbf{B}_k$ be the k-th Bernoulli polynomial (cf. Chapter 2). We had defined the distribution E_k at level N by

$$E_k^{(N)}(x) = N^{k-1} \frac{1}{k} \mathbf{B}_k\left(\left\langle \frac{x}{N} \right\rangle\right).$$

We shall now use

$$N = dp^n,$$

where d is a positive integer prime to the prime number p.

We continue using the notation of the preceding section. An element of $Z = \mathbf{Z}(d) \times \mathbf{Z}_p$ is described by its two components

$$x = (x_0, x_p).$$

Let $c \in \mathbf{Z}(d)^* \times \mathbf{Z}_p^* = \lim \mathbf{Z}(dp^n)^*$. We **define**

$$\boxed{E_{k,c}^{(N)}(x) = E_k^{(N)}(x) - c_p^k E_k^{(N)}(c^{-1}x)}$$

for $x \in \mathbf{Z}(N)$. The multiplication $c^{-1}x$ is defined in $\mathbf{Z}(N)^*$.

Note. In Chapter 2, we took c to be a rational number. This is not necessary, and restricts possible applications too much. When c occurs as a coefficient in Chapter 2, we must use c_p instead of c, i.e. we must use its projection on $\mathbf{Z}_p^*$. When c occurs inside a diamond bracket, then no change is to be made for the present case. For instance, we have

E 1. $\qquad E_{1,c}^{(N)}(x) = \left\langle \frac{x}{N} \right\rangle - c_p\left\langle \frac{c^{-1}x}{N} \right\rangle + \frac{1}{2}(c_p - 1).$

Similarly, formula **E 2** and Theorem 2.2 of Chapter 2 yield the relation

E 2. $\qquad E_{k,c}(x) = x_p^{k-1} E_{1,c}(x)$

symbolically for $x \in Z$. We then obtain the integral representations of the Bernoulli numbers as follows.

$$\frac{1}{k} B_k = \frac{1}{1 - c_p^k} \int_Z x_p^{k-1} \, dE_{1,c}(x),$$

provided only that $c_p^k \neq 1$. Furthermore, if χ is a character of conductor $m = m_\chi$ dividing dp^n for some n, then χ defines in the usual way a function on $\mathbf{Z}(N)$ for $m | N$ by composition

$$\mathbf{Z}(N) \to \mathbf{Z}(m) \xrightarrow{\chi} \mathfrak{o}^*,$$

and χ is defined to be 0 on elements of $\mathbf{Z}(m)$ not prime to m. Then we define

$$\frac{1}{k} B_{k,\chi} = \int_Z \chi \, dE_k.$$

Note. This definition made by taking into account the conductor of χ is more appropriate than that of Chapter 2, §2. There we dealt only with characters of $\mathbf{Z}_p^*$, so it made little difference, only for the trivial character.

More generally, if φ is a locally constant function (step function) on Z, then we can **define**

$$\frac{1}{k} B_{k,\varphi} = \int_Z \varphi \, dE_k.$$

Then

(1) $$\int_Z \varphi(x_0, x_p) x_p^{k-1} \, dE_{1,c}(x) = \frac{1}{k} B_{k,\varphi} - c_p^k \frac{1}{k} B_{k,\varphi \circ c}.$$

In particular, if φ is a character χ, then

$$\int_Z \chi(x) x_p^{k-1} \, dE_{1,c}(x) = (1 - \chi(c) c_p^k) \frac{1}{k} B_{k,\chi}.$$

We define the **p-adic L-function** by the integral

$$L_p(1 - s, \chi) = \frac{-1}{1 - \chi(c) \langle c \rangle_p^s} \int_{Z^*} \chi(a) \langle a \rangle_p^s a_p^{-1} \, dE_{1,c}(a).$$

If the conductor of χ is dp^n for some $n \geq 0$, then the support of the integral is really on the set

$$Z^{**} = \mathbf{Z}(d)^* \times \mathbf{Z}_p^*.$$

Let $\omega = \omega_p$ be the Teichmuller character, and put

$$\chi_k = \chi \omega^{-k}.$$

9

Theorem 2.1. *For every integer* $k \geq 1$ *and character* χ *of conductor* dp^n *with* $n \geq 0$, *we have*

$$L_p(1-k, \chi) = -(1-\chi_k(p)p^{k-1})\frac{1}{k}B_{k,\chi_k}.$$

Proof. We have:

$$-(1-\chi_k(c)c_p^k)L_p(1-k, \chi) = \int_{Z^*} \chi_k(a)a_p^{k-1}\,dE_{1,c}(a).$$

Write

$$\int_{Z^*} = \int_{Z} - \int_{pZ}.$$

Let $N = dp^{n+1}$. Then

$$\int_{pZ} = \lim_{n\to\infty} \sum_{y=0}^{(N/p)-1} \chi_k(p)p^{k-1}\chi_k(y)y^{k-1}E_{1,c}\left(\left\langle\frac{py}{N}\right\rangle\right)$$

$$= \chi_k(p)p^{k-1}\lim_{n\to\infty} \sum_{y=0}^{(N/p)-1} \chi_k(y)y^{k-1}E_{1,c}\left(\left\langle\frac{y}{N/p}\right\rangle\right)$$

$$= \chi_k(p)p^{k-1}(1-\chi_k(c)c_p^k)\frac{1}{k}B_{k,\chi_k}.$$

The theorem follows at once.

We now let

$\theta =$ even character on $\mathbf{Z}(dp)^*$, $\theta \neq 1$, cond $\theta = d$ or dp.
$\chi = \theta\psi$ where ψ is a character on $1 + p\mathbf{Z}_p$.

Then

$$(1-\chi_k(p)p^{k-1})\frac{1}{k}B_{k,\chi_k} = \frac{1}{1-\chi(c)\langle c\rangle_p^k}\int_{Z^{**}} \psi(a)\theta\omega^{-k}(a)a_p^{k-1}\,dE_{1,c}(a)$$

$$= \frac{1}{1-\chi(c)\langle c\rangle_p^k}\int_{Z^{**}} \psi(\langle a_p\rangle)\,d\mu(a)$$

where μ is the measure given by

$$\mu(a) = \theta(a)\omega^{-k}(a)a_p^{k-1}E_{1,c}(a).$$

Therefore by Corollary 1 of Theorem 1.2 we find that

$$(1 - \chi_k(p)p^{k-1})\frac{1}{k}B_{k,\,\chi_k} = \frac{1}{1 - \chi(c)\langle c \rangle_p^k}\,f_{\theta,\,k}(\zeta_\psi - 1)$$

where $f_{\theta,\,k}(X)$ is a power series given mod h_n by Theorem 1.2.

We may use formula **E 1** of Chapter 2, §2 to give the value of μ at intermediate levels, namely

$$\mu_{n+1}(a) = \theta(a)\omega^{-k}(a)a_p^{k-1}\left[\left\langle\frac{a}{dp^{n+1}}\right\rangle - c_p\left\langle\frac{c^{-1}a}{dp^{n+1}}\right\rangle + \tfrac{1}{2}(c_p - 1)\right].$$

Starting with the general formula of Theorem 1.2, we shall derive a slightly simpler expression for the coefficients of $f_{\theta,\,k}$, which can be written in the form

(2) $$f_{\theta,\,k}(X) \equiv \sum_u c_k^{(n)}(u)(1 + X)^{r(u)} \bmod h_n,$$

where

$$c_k^{(n)}(u) = \sum_\eta \theta\omega^{-k}(\eta)(\eta_p u)^{k-1}\left[\left\langle\frac{\eta u}{dp^{n+1}}\right\rangle - c_p\left\langle\frac{c^{-1}\eta u}{dp^{n+1}}\right\rangle + \tfrac{1}{2}(c_p - 1)\right].$$

The sums are taken for $u \in 1 + p\mathbf{Z}_p \bmod p^{n+1}$ and $\eta \in \mathbf{Z}(d)^* \times \boldsymbol{\mu}_{p-1}$. The component η_p is just $\omega(\eta)$. The character $\theta\omega^{-1}$ is odd, and in particular is not trivial. Hence the sum over η times the factor $(c_p - 1)/2$ is equal to 0, and that term can be omitted.

We now select $c \in \mathbf{Z}(d)^* \times \boldsymbol{\mu}_{p-1}$, so that $\langle c \rangle_p = 1$. Furthermore $\chi(c) = \theta(c)$. We can select c such that $\chi(c) \neq 1$. We change variables in the sum over η, with respect to the second term involving $c^{-1}\eta$, letting $\eta \mapsto c\eta$. Then we may combine the sums over both terms, with a factor

$$1 - \chi(c)$$

which cancels $1 - \chi(c)\langle c \rangle_p^k = 1 - \chi(c)$ in front. In other words, we find:

(3) $$c_k^{(n)}(u) = \sum_\eta \theta\omega^{-1}(\eta)u^{k-1}\left\langle\frac{\eta u}{dp^{n+1}}\right\rangle.$$

We are interested in applying Theorem 1.3. In other words, we are interested in proving the Iwasawa conjecture that some coefficient of $f_{\theta,\,k}$ is a

p-unit. Clearly the power u^{k-1} can be disregarded for this purpose. Thus the expressions (3) for the coefficients of the Iwasawa power series give rise to the following criterion.

Theorem 2.2 (Iwasawa congruences). *Let d be an integer ≥ 1 and prime to p. Let θ be an even character $\neq 1$ of conductor d or dp. If no coefficient of $f_{\theta,k}$ is a p-unit, then we have the congruences (independent of k):*

$$\sum_{\eta} \theta\omega^{-1}(\eta)\left\langle \frac{\eta\alpha}{dp^{n+1}} \right\rangle \equiv 0 \bmod \mathfrak{p}$$

for all $\alpha \in \mathbf{Z}_p^*$, and all integers $n \geq 0$.

Proof. We have proved the assertion when α lies in $1 + p\mathbf{Z}_p$. However, for any fixed $\eta_0 \in \boldsymbol{\mu}_{p-1}$ we can make the change of variables

$$\eta \mapsto \eta\eta_0,$$

leading to the congruences as stated above.

Theorem 2.3 (Ferrero–Washington). *For $\theta \neq 1$, not all these congruences are satisfied, and therefore some coefficient of $f_{\theta,k}$ is a p-unit.*

The proof that not all these congruences are satisfied will be given in the next chapter. Here, we first give formulations for these congruences which are more easily dealt with. Then in the next section, we indicate how this result applies to the divisibility of class numbers in the cyclotomic $\mathbf{Z}_p$-extension.

The case $d = 1$. We shall give an alternative version of Iwasawa's congruences adapted for the Ferrero–Washington proof. Write any element $z \in \mathbf{Z}_p^*$ as a series

$$z = z_0 + z_1 p + z_2 p^2 + \cdots$$

with integers z_i satisfying $0 \leq z_i \leq p - 1$. Let

$$s_n(z) = z_0 + z_1 p + \cdots + z_n p^n$$

be the n-th partial sum. In the above congruences, we may replace $\eta\alpha$ by $s_n(\eta\alpha)$, and then omit the brackets giving the representative as a rational number. Furthermore, let us write

$$\theta\omega^{-1} = \omega^v,$$

where v is a positive integer, necessarily odd since we assumed that θ is an even power of the Teichmuller character. Furthermore, $v \not\equiv -1 \bmod p - 1$ because $\theta \neq 1$. Multiplying the congruence by p^{n+1} yields

$$\sum_{\eta \in \mu_{p-1}} s_n(\eta\alpha)\eta^v \equiv 0 \bmod p^{n+2}$$

where v is a positive odd integer, $v \not\equiv -1 \bmod p - 1$.

Now in the p-adic expansion of z we let

$$z_n = t_n(z).$$

We shall express the above congruence in terms of t_n.

Theorem 2.4. *Let $\theta \neq 1$ be an even character of conductor p. Then the Iwasawa congruences imply that there exists an odd integer*

$$v \not\equiv -1 \bmod p - 1$$

such that, for all $\alpha \in \mathbf{Z}_p^*$ and all integers $n \geq 1$ we have

$$2 \sum_{\eta \in \mathcal{R}} t_n(\alpha\eta)\eta^v \equiv (p - 1) \sum_{\eta \in \mathcal{R}} \eta^v \bmod p,$$

where $\mathcal{R}$ is a system of representatives for $\mu_{p-1} \bmod \pm 1$. In particular the congruence class on the left-hand side is independent of α and n.

Proof. We have

$$s_n(\alpha\eta) = s_{n+1}(\alpha\eta) - t_{n+1}(\alpha\eta)p^{n+1}.$$

Furthermore

$$s_{n+1}(\alpha\eta) \equiv \alpha\eta \bmod p^{n+2}$$

and

$$\sum_{\eta \in \mu_{p-1}} \eta^{v+1} = 0$$

because $v \not\equiv -1 \bmod p - 1$. Hence the congruence of the theorem is equivalent to

$$\sum_{\eta} t_{n+1}(\alpha\eta)\eta^v \equiv 0 \bmod p.$$

Since $t_0(\alpha\eta) \equiv \alpha\eta \bmod p$, we always have

$$\sum_{\eta} t_0(\alpha\eta)\eta^v \equiv 0 \bmod p.$$

Finally, since $t_{n+1}(px) = t_n(x)$, we are led to the congruence

$$\sum_\eta t_n(\alpha\eta)\eta^\nu \equiv 0 \bmod p$$

for all n and all α. But since $0 = p + (p-1)p + (p-1)p^2 + \dots$,

$$t_n(-\alpha\eta) = p-1 - t_n(\alpha\eta) \quad \text{for } n \geq 1.$$

Therefore

$$\sum_\eta t_n(\alpha\eta)\eta^\nu = 2 \sum_{\eta \in \mathscr{R}} t_n(\alpha\eta)\eta^\nu - (p-1) \sum_{\eta \in \mathscr{R}} \eta^\nu,$$

thus proving the theorem.

The case $d > 1$.

Theorem 2.5. *Let $\theta \neq 1$ be an even character of conductor d or dp with $d > 1$ prime to p. Let $\theta_1 = \theta\omega^{-1}$. Then the Iwasawa congruences imply that for all $\alpha \in \mathbf{Z}_p^*$ and all $n \geq 0$ we have*

$$\sum_{\eta \in \mathscr{R}} \sum_{i=0}^{d-1} i\theta_1(s_n(\alpha\eta) + ip^{n+1}) \equiv 0 \bmod \mathfrak{p}.$$

Proof. In Theorem 2.2 we may rewrite the congruence in the form

$$\frac{1}{dp^{n+1}} \sum a\theta_1(a) \equiv 0 \bmod \mathfrak{p}$$

where the sum is taken over a prime to dp, such that

$$0 < a < dp^{n+1} \quad \text{and} \quad \langle a \rangle_p \equiv \langle \alpha \rangle_p \bmod p^{n+1}.$$

We can also replace these elements a by elements of the form

$$\eta\alpha + ip^{n+1} \quad \text{with } i = 0, \dots, d-1,$$

and η is some $(p-1)$th root of unity. The sum is then taken over η and i. The sum over i with the factor $\eta\alpha$ is then equal to 0, and we are left only with a sum having ip^{n+1} as a factor. Combining terms with η and $-\eta$, and using the fact that θ_1 is odd yields the desired formula.

§3. Class Numbers as Products of Bernoulli Numbers

We continue to let p be an odd prime. We write $x \sim y$ to mean that $x = yu$ where u is a p-unit. We let:

θ = even character on $\mathbf{Z}(dp)^*$.

ψ = character on $1 + p\mathbf{Z}_p$ of conductor dividing p^{n+1}.

The characters on $\mathbf{Z}(d)^* \times \mathbf{Z}_p^*$ of the same parity as k of conductor dividing dp^{n+1} can be written uniquely in the form

$$\psi\theta\omega^{-k} = \psi\theta_k.$$

For any integer k with $1 \leq k \leq p - 1$, we **define**

$$h_n^{(k)} = p^{n+1} \prod_{\theta \text{ even}} \prod_{\psi} \frac{1}{k} B_{k, \psi\theta_k}.$$

In particular,

$$h_0^{(k)} = p \prod_{\theta \text{ even}} \frac{1}{k} B_{k, \theta_k}.$$

We can simplify these expressions in so far as p-divisibility is concerned. We need a lemma of von Staudt type.

Lemma 1. *Let k be an integer with $1 \leq k \leq p-1$. Then*

$$\frac{1}{k} B_{k, \omega^{-k}} \equiv -\frac{1}{kp} \mod \mathbf{Z}_p.$$

Proof. The proof is entirely similar to that of the Von Staudt congruence, Corollary 2 of Theorem 2.3, Chapter 2, combined with the expression for the Bernoulli number as an integral in Theorem 2.4 of Chapter 2. We leave it to the reader.

Lemma 2. *Let $1 \leq k \leq p-1$. Then*

$$h_0^{(k)} \sim \prod_{\theta \neq 1} \frac{1}{k} B_{k, \theta\omega^{-k}}.$$

Proof. The case when $\theta = 1$ combined with Lemma 1 shows that the factor p in the definition of $h_0^{(k)}$ cancels the pole of order 1 at p of the single term with $\theta = 1$ in the product. What remains is the desired expression.

15

Lemma 3. *Let $1 \leq k \leq p - 1$. Then*

$$h_n^{(k)} \sim h_0^{(k)} \prod_{\theta \neq 1} \prod_{\psi \neq 1} \frac{1}{k} B_{k, \psi\theta\omega^{-k}}.$$

Proof. Write $\chi = \theta\psi$. Then

$$(1 - \chi_k(p)p^{k-1}) \frac{1}{k} B_{k, \chi\omega^{-k}} = \frac{1}{1 - \chi(c)\langle c \rangle_p^k} \int_{Z^*} \chi(a) \langle a \rangle_p^k a_p^{-1} \, dE_{1, c}(a).$$

We distinguish three cases, for the terms in the product defining $h_n^{(k)}$.

If $\theta = 1$ and $\psi = 1$, then we apply Lemma 1. We use one factor of p from p^{n+1} multiplied with

$$\frac{1}{k} B_{k, \omega^{-k}}$$

to find a p-unit.

If $\theta \neq 1$ and $\psi = 1$, then we use Lemma 2 to get the $h_0^{(k)}$ on the right-hand side of the formula to be proved.

The proof of Lemma 3 is concluded by the next lemma.

Lemma 4. *If $\theta = 1$ and $\psi \neq 1$ then*

$$\frac{1}{k} B_{k, \psi\omega^{-k}} \sim \frac{1}{\zeta - 1},$$

where $\gamma = 1 + p$ and $\zeta = \psi(\gamma)$. Furthermore

$$p^n \prod_{\psi \neq 1} \frac{1}{k} B_{k, \psi\omega^{-k}} \sim 1.$$

Proof. We also take $c = 1 + p$. Then

$$1 - \psi(c)\langle c \rangle^k \sim 1 - \zeta \quad \text{and} \quad \prod_{\substack{\zeta p^n = 1 \\ \zeta \neq 1}} (1 - \zeta) = p^n.$$

We note that $\chi(c) = \psi(c)$, and we obtain

$$p^n \prod_{\psi \neq 1} \frac{1}{1 - \psi(c)\langle c \rangle_p^k} \sim 1.$$

Finally we wish to show that

$$\int_{\mathbf{Z}_p^*} \psi(a)\langle a\rangle_p^k a_p^{-1}\, dE_{1,c}(a) \sim 1,$$

i.e. the above integral is a p-unit. Since

$$\psi(a) \equiv 1 \bmod 1 - \zeta \quad \text{and} \quad \langle a \rangle \equiv 1 \bmod p,$$

it suffices to prove that

$$\int_{\mathbf{Z}_p^*} a_p^{-1}\, dE_{1,c}(a)$$

is a p-unit. This is immediate by writing down the first approximation at level p, and concludes the proof.

For each $\theta \neq 1$ we let $\lambda(\theta, k)$ be the linear Iwasawa invariant of the power series $f_{\theta, k}$ in §2, and we let

$$\lambda(k) = \sum_{\theta \neq 1} \lambda(\theta, k).$$

From the Ferrero–Washington theorem and Lemma 3, we then obtain:

Theorem 3.1. *There is a constant c_k such that for all n sufficiently large, we have*

$$\operatorname{ord}_p h_n^{(k)} = \lambda(k)n + c_k.$$

This is merely a special case of Corollary 3 of Theorem 1.2, applied to the Bernoulli distributions, as discussed in §2.

We can then apply the theorem to the class number.

Theorem 3.2. *Let h_n be the class number of $\mathbf{Q}(\mu_{p^{n+1}})$. Then there is a constant c such that for all n sufficiently large, we have*

$$\operatorname{ord}_p h_n^- = \lambda(1)n + c.$$

Proof. The classical class number formula asserts that

$$h_n^- = 2p^{n+1} \prod_{\chi \text{ odd}} -\tfrac{1}{2}B_{1,\chi},$$

so that we can apply Theorem 3.1 with $k = 1$ to conclude the proof.

Theorem 3.3. *Let K be a cyclotomic extension of the rationals (i.e. a subfield of a cyclotomic field). Let K_∞ be the cyclotomic $\mathbf{Z}_p$-extension of K, and let h_n be the class number of K_n. Then there exists a constant c' such that for all n sufficiently large, we have*

$$\operatorname{ord}_p h_n^- = \lambda(1)n + c'.$$

Proof. It is an easy exercise from the class number formula of Chapter 3 to show that the minus part of the class number differs from the product giving $h_n^{(1)}$ only by a finite number of factors. Hence the same estimate holds as in Theorem 3.2.

In Theorem 2.3 of Chapter 12 we shall prove Iwasawa's inequality bounding the order of h_n in terms of the order of h_n^-. We then obtain:

Theorem 3.4. *Notation being as in Theorem 3.3, there exist constants c_1, c_2 (depending on K) such that for all n sufficiently large, we have*

$$\operatorname{ord}_p h_n = c_1 n + c_2.$$

Remark. Iwasawa developed his theory with the point of view that $\mathbf{Z}_p$-extensions are analogous to constant field extensions for curves over finite fields. The formula

$$h_n^- = h_0^- \prod_\zeta f(\xi - 1)$$

is analogous for the function field case of the class number formula. The fact that $\operatorname{ord}_p h_n^-$ is linear in n follows at once from the existence of the Jacobian in the function field case. Kubert–Lang theory suggests the possibility of using the analogous theory in the modular case to analyze the Bernoulli numbers $B_{2,\chi}$ and obtain a bound for the invariant λ in terms of the dimensions of abelian subvarieties of the modular curves.

Appendix by L. Washington: Probabilities

We shall give a heuristic argument which estimates the size of $\lambda_p = \lambda_p(\mathbf{Q}(\mu_p))$. The contribution from λ_p^+ will be ignored, since Vandiver's conjecture says it should be zero. In any case, $\lambda_p^+ \leq \lambda_p^-$, so we could alternatively double our final estimate.

Let $\mathrm{i}(p) = $ index of irregularity $=$ number of Bernoulli numbers $B_2, B_4, \ldots, B_{p-3}$ which are divisible by p. The idea will be to show that usually

$$\lambda_p = \mathrm{i}(p),$$

and that one should expect

$$\lambda_p \leq i(p) + 1$$

for all but a finite number of p.

There are $(p - 3)/2$ relevant power series. We assume that each coefficient is random mod p, and that these coefficients behave independently of each other. The first coefficients of these power series correspond to the Bernoulli numbers in such a way that a first coefficient is divisible by p exactly when the corresponding Bernoulli number is divisible by p. The numerical evidence bears out the assumption that the Bernoulli numbers are random mod p. However, we are also assuming that the higher coefficients are random and independent of each other. This is a more dangerous assumption, and I know of no supporting numerical evidence.

Suppose $\lambda_p \geq i(p) + 2$. Then we have two cases.

Case 1. *Some power series has its first three coefficients divisible by p.* The probability that at least one of the first three coefficients for a given power series is not divisible by p is $1 - 1/p^3$. The probability that for all $(p-3)/2$ power series we have one of the first three coefficients not divisible by p is

$$\left(1 - \frac{1}{p^3}\right)^{(p-3)/2}.$$

Therefore the probability that at least one power series has its first three coefficients divisible by p is

$$1 - \left(1 - \frac{1}{p^3}\right)^{(p-3)/2} = O\left(\frac{1}{p^2}\right).$$

The expected number of times this should happen is therefore finite, since $\sum 1/p^2 < \infty$.

Case 2. *At least two different series have their first two coefficients divisible by p.* Reasoning as in Case 1, we see that the probability that none of the power series has both of the first two coefficients divisible by p is

$$\left(1 - \frac{1}{p^2}\right)^{(p-3)/2}.$$

The probability that exactly one has its first two coefficients divisible by p is

$$\binom{(p-3)/2}{1}\left(1 - \frac{1}{p^2}\right)^{((p-3)/2)-1}\left(\frac{1}{p^2}\right).$$

19

Therefore, the probability that at least two power series have their first two coefficients divisible by p is

$$1 - \left[\left(1 - \frac{1}{p^2}\right)^{((p-3)/2)} + \binom{(p-3)/2}{1}\left(1 - \frac{1}{p^2}\right)^{((p-3)/2)-1}\left(\frac{1}{p^2}\right)\right] = O\left(\frac{1}{p^2}\right).$$

So again one expects only finitely many occurrences.

We therefore expect

$$i(p) \le \lambda_p \le i(p) + 1$$

for all but finitely many p. Therefore estimating λ_p is equivalent to estimating $i(p)$, which we shall do.

However, first we shall show that usually one should expect $\lambda_p = i(p)$, as was the case in Wagstaff's calculations for $p \le 125{,}000$.

If $\lambda_p \ge i(p) + 1$, then at least one power series has its first two coefficients divisible by p. The probability is

$$1 - \left(1 - \frac{1}{p^2}\right)^{((p-3)/2)} = \frac{1}{2p} + O\left(\frac{1}{p^2}\right).$$

Therefore the number of expected occurrences of $\lambda_p \ge i(p)+1$ for $p \le x$ should be

$$\sum_{p \le x} \frac{1}{2p} \sim \frac{1}{2} \log \log x.$$

Since $\frac{1}{2} \log \log (125{,}000) \sim 1.2$, it is not very surprising that $\lambda_p = i(p)$ for $p < 125{,}000$. In fact, one might expect to search rather far before finding a counterexample. A reasonable bound might be 10^{24} since $\frac{1}{2} \log \log 10^{24} \sim 2$. Also note that the fact that

$$1 < \frac{1}{2} \log \log 125{,}000$$

is really caused by the first few primes. If one considers, for example,

$$\sum_{30 < p < x} \frac{1}{2p},$$

then the expected number is much less than 1. Starting the sum at $p = 31$ is perhaps justified by the fact that the early Bernoulli numbers, etc., are too small to be random mod p. In fact, even though 39 % of primes are irregular, 37 is the first one.

We now estimate i(p). The probability that i(p) = i is

$$\binom{(p-3)/2}{i}\left(1 - \frac{1}{p}\right)^{((p-3)/2)-i}\left(\frac{1}{p}\right)^i \xrightarrow[\text{as } p \to \infty]{} e^{-1/2}\frac{(\frac{1}{2})^i}{i!}.$$

The right-hand side is the Poisson distribution. The probability is as stated because i of the Bernoulli numbers are divisible by p, each with probability $1/p$. There are $(p - 3)/2 - i$ of them not divisible by p, each with probability $1 - 1/p$. Finally, there are

$$\binom{(p-3)/2}{i}$$

ways of choosing the i Bernoulli numbers which are divisible by p.

For $i = 0$, we obtain the "result" that the fraction of regular primes is $e^{-1/2} \sim 61\%$.

The number of occurrences of i(p) = i for $p \leq x$ should be approximately

$$\frac{x}{\log x}e^{-1/2}\frac{(\frac{1}{2})^i}{i!}.$$

We should therefore expect the first occurrence to be when

$$\frac{x}{\log x}e^{-1/2}\frac{(\frac{1}{2})^i}{i!} \sim 1.$$

Taking logarithms and ignoring lower order terms, we find, with the help of Stirling's formula:

$$\log x \sim i \log i \quad \text{so} \quad \log \log x \sim \log i \quad \text{and} \quad \frac{\log x}{\log \log x} \sim i.$$

Since x was the first occurrence of i, we obtain approximately

$$i(p) \leq \frac{\log p}{\log \log p} \quad \text{therefore} \quad \lambda_p \leq \frac{\log p}{\log \log p}.$$

21

For $p \sim 125,000$, this yields $\lambda_p \leq 4.8$ which is close to the truth, namely $\lambda_p \leq 5$. Of course, most of the time λ_p will be much less than this bound: 61% of the time we should have $\lambda_p = 0$. The "average value" of λ_p is

$$\lim_{x \to \infty} \frac{\sum_{p \leq x} \lambda_p}{\sum_{p \leq x} 1} = \lim \left(\frac{\sum i(p)}{\sum 1} + \frac{O(\log \log x)}{x/\log x} \right)$$

$$= \lim \frac{\sum i(p)}{\sum 1}$$

$$= \sum_{i=0}^{\infty} (i) \, (\text{probability that } i(p) = i)$$

$$= \sum_{i=0}^{\infty} i e^{-1/2} \frac{(\tfrac{1}{2})^i}{i!}$$

$$= \tfrac{1}{2}.$$

Bibliography. D. H. Lehmer seems to have been the first to use probability arguments such as the above, since he mentioned that $1 - e^{-1/2} = 39\%$ of primes are irregular in [Leh]. Later, Siegel published a probability argument giving this result in [Si 2]. Numerical evidence appears in Johnson [Jo] and Wagstaff [Wag]. Kummer (last page of vol. I of his collected works) claimed that a simple probability argument yields the ratio of irregular to regular primes is $\tfrac{1}{2}$, but it appears he was mistaken.

§4. Divisibility by l Prime to p: Washington's Theorem

Theorem 4.1. *Let l, p be distinct primes. If $p \neq 2$, let $q = p$ and if $p = 2$, let $q = 4$. Let χ be an odd Dirichlet character of conductor dividing dq with $(d, p) = 1$. If ψ is a character on $1 + q\mathbf{Z}_p$ with conductor p^{n+1} sufficiently large (depending on l and χ), then the Bernoulli number $\tfrac{1}{2}B_{1, \chi\psi}$ is an l-unit.*

Before proving the theorem, we give its application to the class numbers of cyclotomic fields.

Theorem 4.2. *Let K be a cyclotomic extension of $\mathbf{Q}$. Let K_∞ be the cyclotomic $\mathbf{Z}_p$-extension of K. Let l be a prime $\neq p$. Then*

$$\text{ord}_l |C(K_n)|$$

is bounded.

Proof. By lemma 2, §1 of Chapter 13 it suffices to prove the theorem when $K = \mathbf{Q}(\mu_{dq})$ for some positive integer d not divisible by p. Furthermore, we may also adjoin an l-th root of unity to the ground field, and thus assume without loss of generality that d is divisible by l. Theorem 3.2 of Chapter 13 then shows that it suffices to prove that

$$\operatorname{ord}_l h_n^-$$

is bounded. But we have the formula

$$h_n^- = Q_n w_n \prod_\chi \prod_\psi - \tfrac{1}{2} B_{1, \chi\psi},$$

where χ ranges over all odd characters of $\mathbf{Z}(dq)^*$ and ψ ranges over all characters of $1 + q\mathbf{Z}_p$ of conductor dividing p^{n+1}. The factor Q_n is Hasse's index, equal to 1 or 2, and w_n is l-bounded. Hence we may apply Theorem 4.1 to conclude the proof.

In the rest of this section, we reduce Theorem 4.1 to congruences similar to those which we have already met. To avoid using notation involving 4 in case $p = 2$, we assume that p is odd. Actually, the case $p = 2$ is easier and was solved by Washington before the general case.

Let $\psi = \psi_n$ have conductor p^{n+1}. Let

$$F_n = \mathbf{Q}(\chi, \psi) = \mathbf{Q}(\chi, \boldsymbol{\mu}_{p^n})$$

be the field obtained by adjoining the values of χ and ψ_n to $\mathbf{Q}$. Then $B_{1, \chi\psi}$ belongs to F_n. Let $T_{n, m}$ be the trace from F_n to F_m. Let $m_0 \geq 1$ be a positive integer such that if

$$n > m \geq m_0$$

then $F_n \neq F_m$. Finally, let $\mathcal{R}$ be a set of representatives of the group of $(p-1)$-th roots of unity in $\mathbf{Z}_p$, modulo ± 1.

Given the prime $l \neq p$, there exist only a finite number of prime ideals in $\mathbf{Q}(\boldsymbol{\mu}^{(p)})$ lying above l, as is seen immediately from the structure of the decomposition group. It follows at once that if

$$F_\infty = \bigcup F_n,$$

then there is only a finite number of prime ideals in F_∞ lying above l. Let $\mathfrak{L}$ be one of these primes. We shall now indicate how to prove that given χ, for all ψ with sufficiently large conductor p^{n+1}, we have

$$\tfrac{1}{2} B_{1, \chi\psi} \not\equiv 0 \bmod \mathfrak{L}.$$

We can choose an integer m such that all primes of F_m above l remain prime in F_∞, and we take $m \geq m_0$. Let $n \geq m$.

Suppose that $\frac{1}{2}B_{1,\chi\psi} \equiv 0 \bmod \mathfrak{L}$. Since $\mathfrak{L}$ is the only prime of F_∞ lying above its restriction to F_{m0}, we conclude that

$$T_{n,m}(\tfrac{1}{2}\psi(\alpha)^{-1}B_{1,\chi\psi}) \equiv 0 \bmod \mathfrak{L}$$

for all $\alpha \in \mathbf{Z}_p^*$.

We shall now transform this congruence into more explicit terms.

Lemma 4.3. If $\frac{1}{2}B_{1,\chi\psi} \equiv 0 \bmod \mathfrak{L}$ for infinitely many ψ (so of arbitrarily large conductor p^{n+1}), then there exist infinitely many n such that for such ψ and all $\alpha \in \mathbf{Z}_p^*$ we have

$$\frac{\psi(\alpha)^{-1}}{d} \sum_{\eta \in \mathscr{R}} \sum_{r=0}^{dp^m-1} r\chi\psi(s_{n-m}(\alpha\eta) + rp^{n-m+1}) \equiv 0 \bmod \mathfrak{L}.$$

Proof. Abbreviate $T = T_{n,m}$. From the irreducible equation of a p-power root of unity, we see at once that $T(\varepsilon) = 0$ for any p-power root of unity ε which does not lie in F_m. Thus if $\beta \in \mathbf{Z}_p^*$ and we write

$$\beta = \omega(\beta)\langle\beta\rangle_p$$

where ω is the Teichmuller character and $\langle\beta\rangle_p \equiv 1 \bmod p$, we get:

$$T(\psi(\beta)) \neq 0 \Leftrightarrow \psi(\beta)^{p^m} = 1$$
$$\Leftrightarrow \langle\beta\rangle_p^{p^m} \equiv 1 \bmod p^{n+1}$$
$$\Leftrightarrow \langle\beta\rangle_p \equiv 1 \bmod p^{n-m+1}.$$

Consequently we find

$$T(\tfrac{1}{2}\psi(\alpha)^{-1}B_{1,\chi\psi}) = \tfrac{1}{2}p^{n-m} \sum \chi(a)\psi(a\alpha^{-1})\frac{a}{dp^{n+1}}$$

where the sum is taken for

$$0 < a < dp^{n+1} \quad \text{and} \quad \langle a\rangle_p \equiv \langle\alpha\rangle_p \bmod p^{n-m+1}$$

This can be rewritten

$$T(\tfrac{1}{2}\psi(\alpha)^{-1}B_{1,\chi\psi}) = \frac{\psi(\alpha)^{-1}}{2\,dp^{m+1}} \sum_{\eta \in \mathscr{R}} S(\eta, \alpha)$$

where

$$S(\eta, \alpha) = \sum_{a \equiv \pm \eta\alpha(p^{n-m+1})} \chi(a)\psi(a)a.$$

Note that given η, elements a satisfying

$$0 < a < dp^{n+1} \quad \text{and} \quad a \equiv \eta\alpha \bmod p^{n-m+1},$$

can be paired with elements a' such that

$$0 < a' < dp^{n+1} \quad \text{and} \quad a' \equiv -\eta\alpha \bmod p^{n-m+1}$$

of the form

$$a' = dp^{n+1} - a.$$

Since χ is odd and ψ is even, we therefore find

$$S(\eta, \alpha) = \sum_{a \equiv \eta\alpha(p^{n-m+1})} [2\chi(a)\psi(a)a - \chi(a)\psi(a) \, dp^{n+1}].$$

But the integers a satisfying the above conditions are precisely those of the form

$$s_{n-m}(\alpha\eta) + rp^{n-m+1} \quad \text{with } 0 \leq r \leq dp^m - 1.$$

This concludes the proof of the lemma.

To prove Theorem 4.1, we still have to deal with the possibility that $\frac{1}{2}B_{1,\chi\psi}$ has a pole at $\mathfrak{L}$. However, we note that $\mathfrak{L}$ has finite ramification index over l. Consequently for each χ there is only a finite number of ψ such that

$$\tfrac{1}{2}B_{1,\chi\psi}$$

has a pole at $\mathfrak{L}$, because such terms contribute negative l-order to the class number h_n^-, which can be cancelled only by a finite number of other factors in light of the congruence

$$\tfrac{1}{2}B_{1,\chi\psi} \not\equiv 0 \bmod \mathfrak{L}$$

proved for all but a finite number of ψ. This concludes the proof of the assertion in Theorem 4.1 that in fact, all but a finite number of $\frac{1}{2}B_{1,\chi\psi}$ are l-units.

11 The Ferrero–Washington Theorems

In this chapter we prove that the Iwasawa congruences cannot be satisfied, thus giving a bound for the divisibility of Bernoulli numbers with characters, and hence a bound for the divisibility of the corresponding class numbers with respect to certain primes.

The proofs closely follow Ferrero–Washington [Fe–W] and Washington [Wa 2], except that Gillard [Gi 2] gave a simplification which we take into account.

§1. Basic Lemma and Applications

The impossibility of the congruences derived in the preceding chapter will follow from the next lemma, valid for some choice of representatives $\mathscr{R}$ for μ_{p-1} (mod ± 1).

Lemma 1.1. *Let d, m be positive integers with d prime to p. For all n sufficiently large, there exists $\alpha_1, \alpha_2 \in \mathbf{Z}_p$, with $\alpha_1, \alpha_2 \equiv 1 \bmod p^m$, and an element $\eta_0 \in \mathscr{R}$, having the following properties.*

(i) $s_n(\alpha_1\eta) = s_{n-m}(\alpha_1\eta) \equiv 0 \bmod d$ *for all* $\eta \in \mathscr{R}$;
(ii) $s_n(\alpha_2\eta) = s_{n-m}(\alpha_2\eta) \equiv 0 \bmod d$ *for all* $\eta \neq \eta_0$;
(iii) $s_n(\alpha_2\eta_0) = s_{n-m}(\alpha_2\eta_0) + p^{n-m+1} \equiv 0 \bmod d.$

The proof of this lemma will be given in the next sections.

Although Ferrero–Washington used similar lemmas for their theorems, Gillard [Gi 2] observed that the above single statement suffices in all cases. We now show its applications in each of the three cases under consideration.

The congruences of Theorem 2.4 of Chapter 10. We recall that these congruences are:

(1) $$\sum_{\eta \in \mathcal{R}} t_n(\alpha\eta)\eta^{\nu} \bmod p \text{ is independent of } \alpha \text{ and } n.$$

In this case, we can take $m = d = 1$, and the congruences mod d become irrelevant, so that

$$t_n(\alpha_1\eta) = t_n(\alpha_2\eta) = 0 \quad \text{for all } \eta \neq \eta_0;$$

$$t_n(\alpha_1\eta_0) = 0 \quad \text{and} \quad t_n(\alpha_2\eta_0) = 1.$$

Subtracting the corresponding expressions in the congruences yields the contradiction.

The congruences of Theorem 2.5 of Chapter 10. We recall that these congruences are:

(2) $$\sum_{\eta \in \mathcal{R}} \sum_{i=0}^{d-1} i\chi(s_n(\alpha\eta) + ip^{n+1}) \equiv 0 \bmod \mathfrak{p}.$$

The character χ is odd, of conductor d or dp, and $(d, p) = 1$, while $d \neq 1$. We take $m = 2$ in Lemma 1.1. Then we obtain for $\eta \neq \eta_0$:

$$s_n(\alpha_1\eta) \equiv \eta \equiv s_n(\alpha_2\eta) \bmod p$$

$$s_n(\alpha_1\eta) \equiv 0 \equiv s_n(\alpha_2\eta) \bmod d,$$

and hence $s_n(\alpha_1\eta) \equiv s_n(\alpha_2\eta) \bmod dp$. Since χ has conductor d or dp, we get for $i = 0, \ldots, d-1$:

$$\chi(s_n(\alpha_1\eta) + ip^{n+1}) = \chi(s_n(\alpha_2\eta) + ip^{n+1}).$$

Similarly,

$$s_n(\alpha_2\eta_0) \equiv s_n(\alpha_1\eta_0) - p^{n+1} \bmod dp,$$

Let $a = s_n(\alpha_1\eta_0)$. Then the congruence (2) yields

$$\sum_{i=0}^{d-1} i\chi(a+ip^{n+1}) \equiv \sum_{i=0}^{d-1} i\chi(a+(i-1)p^{n+1}) \bmod \mathfrak{p}.$$

But

$$\sum_{i=0}^{d-1} \chi(a+ip^{n+1}) = 0$$

because the conductor of χ does not divide p^{n+1}. Hence

$$\sum_{i=0}^{d-1}(i+1)\chi(a+ip^{n+1}) \equiv \sum_{i=0}^{d-1} i\chi(a+(i-1)p^{n+1}) \bmod \mathfrak{p}.$$

Subtracting yields

$$d\chi(a+(d-1)p^{n+1}) \equiv 0 \bmod \mathfrak{p}.$$

This is a contradiction because $a+(d-1)p^{n+1}$ is prime to dp. Indeed, it is obviously prime to p because it is $\equiv \eta_0 \bmod p$, and on the other hand, since $a \equiv 0 \bmod d$, we get

$$a+(d-1)p^{n+1} \equiv -p^{n+1} \bmod d.$$

This concludes the proof.

The congruences of Lemma 4.3 of Chapter 10. We recall that these congruences are:

$$(3) \qquad \frac{1}{d}\sum_{\eta \in \mathscr{R}}\sum_{r=0}^{dp^m-1} r\chi\psi(s_{n-m}(\alpha\eta)+rp^{n-m+1}) \equiv 0 \bmod \mathfrak{L}.$$

For $\alpha = \alpha_1$ or $\alpha = \alpha_2$, we get for all $\eta \in \mathscr{R}$:

$$\alpha^{-1}(s_n(\alpha\eta)+rp^{n-m+1}) \equiv \eta+rp^{n-m+1} \bmod p^{n+1}.$$

Let $a = s_n(\alpha_1\eta_0)$. Arguing as in the preceding case, we get

$$\frac{1}{d}\sum_{i=0}^{dp^m-1} i\chi(a+ip^{n-m+1})\psi(\eta_0+ip^{n-m+1})$$

$$\equiv \frac{1}{d}\sum_{i=0}^{dp^m-1} i\chi(a+(i-1)p^{n-m+1})\psi(\eta_0+(i-1)p^{n-m+1}) \bmod \mathfrak{L}.$$

Again changing i to $i+1$ as in the preceding case, we get

$$\frac{1}{d}dp^m\chi(a+(dp^m-1)p^{n-m+1})\psi(\eta_0+(dp^m-1)p^{n-m+1}) \equiv 0 \bmod \mathfrak{L}.$$

But $a+(dp^m-1)p^{n-m+1}$ is prime to dp and $\eta_0+(dp^m-1)p^{n-m+1}$ is prime to p, so we get the desired contradiction.

§2. Equidistribution and Normal Families

We recall some facts about equidistribution of sequences on $\mathbf{R}/\mathbf{Z}$. Let $\mathscr{F}$ be a family of Riemann integrable functions of $\mathbf{R}/\mathbf{Z}$. Let $\mathbf{C}\langle\mathscr{F}\rangle$ be the vector space generated by $\mathscr{F}$. We say that $\mathbf{C}\langle\mathscr{F}\rangle$ is **Riemann dense** (in the space of all Riemann integrable functions) if given ε and given a real Riemann integrable function g, there exist real functions $f_1, f_2 \in \mathbf{C}\langle\mathscr{F}\rangle$ such that

$$f_1 \le g \le f_2$$

and

$$\int_{\mathbf{R}/\mathbf{Z}} (f_2 - f_1) < \varepsilon.$$

(In other words, we can approximate g above and below by functions from $\mathbf{C}\langle\mathscr{F}\rangle$.)

Let $\{y_n\}$ be a sequence of elements in $\mathbf{R}/\mathbf{Z}$. Consider the condition:

EQU. *Let $\mathscr{F}$ be a family of (complex valued) Riemann-integrable functions on $\mathbf{R}/\mathbf{Z}$ such that the vector space generated by $\mathscr{F}$ is Riemann dense. Then for every function f in $\mathscr{F}$, we have*

$$\int_{\mathbf{R}/\mathbf{Z}} f(x)\, dx = \lim_{N \to \infty} \frac{1}{N}(f(y_1) + \cdots + f(y_N)).$$

By a three epsilon argument, one sees that if the sequence satisfies **EQU** for one family $\mathscr{F}$, then it satisfies **EQU** for every such family. If that is the case, we say that the sequence is **equidistributed**, or **uniformly distributed**. Examples of such families which we shall use are as follows.

The most classical family is the family of characteristic functions of intervals $[a, b)$ contained in $[0, 1)$. Then equidistribution means that the density of n such that y_n lies in $[a, b)$ exists and is equal to the measure of $[a, b)$. In other words,

$$\lim_{N \to \infty} \frac{1}{N}(\text{number of } n \le N \text{ such that } y_n \in [a, b)) = b - a.$$

In the application we deal with an even more restricted family, when the end points a, b of the intervals satisfy additional restrictions (rational numbers whose denominators are powers of a prime p), but the sequence still satisfies **EQU** with respect to this family.

We shall also deal with the family of characters on $\mathbf{R}/\mathbf{Z}$, i.e., functions of type $e^{2\pi i m y}$, which satisfies **EQU**. Then **EQU** is known as **Weyl's criterion**.

The above criteria apply mutatis mutandis to r-space $\mathbf{R}^r/\mathbf{Z}^r$, using cubic boxes instead of intervals.

Let $\{\beta_1, \ldots, \beta_r\}$ be a family of p-adic integers. The following two conditions are equivalent, and define what it means for this family to be **normal**.

NOR 1. *For every positive integer k and every $r \times k$ matrix (c_{ij}) of integers with $0 \leq c_{ij} \leq p-1$, there exists $n \geq -1$ such that*

$$t_{n+j}(\beta_i) = c_{ij}$$

for $i = 1, \ldots, r$ and $j = 1, \ldots, k$, and in fact the asymptotic density of such n is p^{-rk}.

The condition means that every possible block of coefficients appears in the p-adic expansions of $\beta_1, \ldots, \beta_r$ with the expected frequency.

NOR 2. *The sequence*

$$\left\{ \left(\frac{1}{p^{n+1}} s_n(\beta_1), \ldots, \frac{1}{p^{n+1}} s_n(\beta_r) \right) \right\}, \qquad n = 1, 2, \ldots$$

is uniformly distributed mod $\mathbf{Z}^r$.

We shall now prove that these two conditions are equivalent.

Let $\beta \in \mathbf{Z}_p$ and let $C = (c_1, \ldots, c_k)$ be a k-tuple of integers with

$$0 \leq c_i \leq p-1.$$

Denote the principal part

$$\mathrm{Pr}_k(C) = \frac{c_1}{p^k} + \cdots + \frac{c_k}{p}.$$

Let $I_k(C)$ be the interval of real numbers $[a, a + 1/p^k)$ where $a = \mathrm{Pr}_k(C)$. Write

$$s_n(\beta) - s_{n-k}(\beta) = z_{n-k+1} p^{n-k+1} + \cdots + z_n p^n.$$

Define the principal part

$$\mathrm{Pr}_{n,k}(\beta) = \frac{1}{p^{n+1}} (s_n(\beta) - s_{n-k}(\beta)) = \frac{z_{n-k+1}}{p^k} + \cdots + \frac{z_n}{p}.$$

Then one verifies at once that

$$\Pr_{n,k}(\beta) = \Pr_k(C) \quad \text{if and only if} \quad \frac{1}{p^{n+1}} s_n(\beta) \in I_k(C).$$

Applying this criterion to an r-tuple $\beta_1, \ldots, \beta_r$, we see that if the sequence in **NOR 2** is equidistributed, then $\{\beta_1, \ldots, \beta_r\}$ satisfies **NOR 1**. Conversely, we also see from the above criterion that if $\{\beta_1, \ldots, \beta_r\}$ satisfies **NOR 1**, then that sequence is equidistributed over intervals of type $[a, a + 1/p^k)$, with $a = \Pr_k(C)$ as above. We can then apply **EQU**, with the family of characteristic functions of such intervals.

Remark. As a special case, we also see that $t_n(\beta)$ depends only on the interval $[a/p, (a+1)/p)$ in which $p^{-(n+1)}s_n(\beta)$ lies. This remark will be used in the applications. For instance if $p^{-(n+1)}s_n(\beta)$ lies in the interval $[0, 1/p)$, then $t_n(\beta) = 0$. If it lies in the interval $[1/p, 2/p)$ then $t_n(\beta) = 1$, and so forth.

We use Haar measure on $\mathbf{Z}_p$ (normalized to have total measure 1), and the expression "almost all" refers to all elements except on a set of measure 0 for that measure.

Lemma 2.1. *Let* $\{\beta_1, \ldots, \beta_r\}$ *be elements of* $\mathbf{Z}_p$ *which are linearly independent over the rationals. Then for almost all* $\alpha \in \mathbf{Z}_p$ *the family* $\{\alpha\beta_1, \ldots, \alpha\beta_r\}$ *is normal.*

Proof. By Weyl's criterion, we must show that for every r-tuple of integers $(a_1, \ldots, a_r)$ not all 0, and almost all α, we have

$$\lim_{N \to \infty} \frac{1}{N} \sum_{n=1}^{N} e\left(\sum_{i=1}^{r} \frac{1}{p^{n+1}} s_n(\alpha\beta_i)a_i \right) = 0,$$

where $e(x) = e^{2\pi i x}$. (For each r-tuple we exclude a set of measure 0, but there are only countably many r-tuples.) Let

$$\beta = \sum_{i=1}^{r} a_i \beta_i.$$

Since

$$s_n(\alpha\beta) \equiv \alpha\beta \equiv \sum_{i=1}^{r} s_n(\alpha\beta_i)a_i \bmod p^{n+1}$$

31

it suffices to show that

$$\lim_{N \to \infty} \frac{1}{N} \sum_{n=1}^{N} e\left(\frac{1}{p^{n+1}} S_n(\alpha\beta)\right) = 0$$

for almost all α. Let

$$S(N, \alpha) = \frac{1}{N} \sum_{n=1}^{N} e\left(\frac{1}{p^{n+1}} S_n(\alpha\beta)\right).$$

Then writing $|S(N, \alpha)|^2 = S(N, \alpha)\overline{S(N, \alpha)}$ we find:

$$\int_{\mathbf{Z}_p} |S(N, \alpha)|^2 \, d\alpha = \frac{1}{N} + \frac{1}{N^2} \sum_{m \neq n} \int_{\mathbf{Z}_p} (\text{cross terms}) \, d\alpha$$

$$= \frac{1}{N},$$

because the integral of the cross terms is equal to 0, since the integral of a non-trivial character over the group $\mathbf{Z}_p$ is equal to 0. (In this case, the integral is a sum of roots of unity.) Thus we obtain

$$\sum_{m=1}^{\infty} \int |S(m^2, \alpha)|^2 \, d\alpha = \sum_{m=1}^{\infty} \frac{1}{m^2} < \infty.$$

By Fubini's theorem, we can put the summation sign on the left inside the integral, and thus conclude that

$$\lim_{m \to \infty} S(m^2, \alpha) = 0$$

for almost all α.

For arbitrary N, choose m such that $m^2 \leq N < (m+1)^2$. Trivial estimates show that

$$|S(N, \alpha)| \leq |S(m^2, \alpha)| + \frac{2m}{N} \to 0 \quad \text{as } N \to \infty.$$

This concludes the proof.

32

§3. An Approximation Lemma

Lemma 3.1. *Let* $\{\beta_1, \ldots, \beta_r\}$ *be p-adic integers, linearly independent over the rationals. Suppose we are given* $\varepsilon > 0$; *an integer* $m > 0$; *an integer* d *with* $(p, d) = 1$; *real numbers* $x_1, \ldots, x_r \in (0, 1)$. *Then for all* n *sufficiently large, there exists* $\alpha \in \mathbf{Z}_p$ *satisfying:*

(i) $\alpha \equiv 1 \bmod p^m$;

(ii) $|p^{-(n+1)} s_n(\alpha \beta_i) - x_i| < \varepsilon$ *for* $i = 1, \ldots, r$;

(iii) $s_n(\alpha \beta_i) \equiv 0 \bmod d$ *for* $i = 1, \ldots, r$.

Proof. We use vector notation and put $x = (x_1, \ldots, x_r)$, $\beta = (\beta_1, \ldots, \beta_r)$. We let $\| \ \|$ be the sup norm on the torus $\mathbf{R}^r/\mathbf{Z}^r$. For each n we define the residue

$$\mathrm{res}_n(\beta) = p^{-(n+1)}(s_n(\beta_1), \ldots, s_n(\beta_r)).$$

We may assume that ε is so small that the intervals $[x_i - \varepsilon, x_i + \varepsilon]$ are contained in the open interval $(0, 1)$ for all i. Select N sufficiently large so that $1/N < \varepsilon/2d$. By Lemma 2.1, for each r-tuple

$$k = (k_1, \ldots, k_r)$$

of integers k_i with $0 \le k_i \le N-1$ we can find a p-adic integer α_k and an integer n_k such that

$$\left\| \mathrm{res}_{n_k}(\alpha_k \beta) - \frac{k}{N} \right\| < \varepsilon/2d.$$

Let $n_0 = m + \max_k n_k$ and let $n \ge n_0$.

There exists some k such that

$$\left\| \frac{x}{d} - \mathrm{res}_n\left(\frac{\beta}{d}\right) - \frac{k}{N} \right\| < \varepsilon/2d.$$

Let $\alpha' = (1/d) + p^{n-n_k}\alpha_k$. Then $\alpha' \equiv 1/d \bmod p^m$, and

$$\left\| \mathrm{res}_n\left(\frac{\beta}{d}\right) + \mathrm{res}_{n_k}(\alpha_k \beta) - \mathrm{res}_n(\alpha' \beta) \right\| = 0.$$

Hence the above inequalities yield

$$\left\| \frac{x}{d} - \mathrm{res}_n(\alpha' \beta) \right\| < 2\varepsilon/2d = \varepsilon/d.$$

It follows that

$$p^{-(n+1)}s_n(\alpha'\beta_i) \in \left[\frac{x_i-\varepsilon}{d}, \frac{x_i+\varepsilon}{d}\right]$$

for all $i = 1, \ldots, r$.

Finally we let $\alpha = d\alpha'$. Then $s_n(\alpha\beta_i) = ds_n(\alpha'\beta_i)$, and α satisfies the required conditions.

The above proof, considerably simpler than the original proof, is taken from the Bourbaki report by Oesterle (Bourbaki Seminar, February 1979).

§4. Proof of the Basic Lemma

This section contains the proof of Lemma 1.1.

Let η^1 be a primitive $(p-1)$th root of unity, and let its powers be

$$\eta^j \quad \text{for } j = 1, \ldots, R \quad \text{where} \quad R = \frac{p-1}{2}.$$

Then these powers η^j represent the elements of $\mu_{p-1}/\pm 1$.

We shall write $\eta_1, \ldots, \eta_r$ for $\pm\eta^1, \ldots, \pm\eta^r$, with any choice of sign, and $r = \phi(p-1)$. We can express

$$\eta^j = \eta_j = \sum_{i=1}^r a_{ji}\eta_i \quad \text{for } j = r+1, \ldots, R,$$

with integral coefficients a_{ji}. Thus we obtain an $(R-r) \times r$ matrix

$$A = (a_{ji}), \qquad (r+1 \le j \le R; \ 1 \le i \le r).$$

We let $x_1, \ldots, x_r$ be real numbers, and we then let

$$x_j = \sum_{i=1}^r a_{ji}x_i \quad \text{for } j = r+1, \ldots, R.$$

Observe that changing the signs of $\eta_1, \ldots, \eta_r$ amounts to changing the signs of the columns in the matrix A. Changing the signs of $\eta_{r+1}, \ldots, \eta_R$ amounts to changing the signs of the rows. Such changes of signs will be called admissible.

Since η_i/η_j is not rational for $i \ne j$, it follows that in each row, there are at least two non-zero elements. Furthermore, it is clear that no two rows of the matrix A are equal. We now prove:

Lemma 4.1. *Let $A = (a_{ji})$ be a real matrix such that no row equals $+$ or $-$ another, and in every row there are at least two non-zero elements. After an admissible change of sign, we can find a vector $(x_1, \ldots, x_r) = {}^t X^{(r)} \in \mathbf{R}^r$ such that, if we put*

$$X^{(R-r)} = A X^{(r)} \quad and \quad X^{(R-r)} = {}^t(x_{r+1}, \ldots, x_R),$$

then:

(i) *We have $x_j > 0$ for $j = 1, \ldots, R$;*
(ii) *We have $x_{j'} \neq x_j$, for all $j \neq j'$.*

Proof. In R-space, we consider the conditions:

$x_j = 0$ for some $j = 1, \ldots, R$;
$x_j - x_{j'} = 0$ or $x_j + x_{j'} = 0$ for some pair (j, j') with $j \neq j'$.

Each such condition defines a hyperplane. We want some $X^{(r)} \in \mathbf{R}^r$ such that

$$(X^{(r)}, A X^{(r)})$$

does not lie in the union of these hyperplanes. Let V be the vector space of all vectors $(X^{(r)}, A X^{(r)})$, i.e. the graph of the linear map represented by A. Then V is not contained in any one of the above hyperplanes because of the two assumptions on the matrix A. Hence V is not contained in the finite union of these hyperplanes. Let V' be the complement of these hyperplanes in V, and let V'_0 be the projection of V' on the first r coordinates. Take $(x_1, \ldots, x_r)$ in the positive 2^r-quadrant intersected with V'_0, and let $(x_1, \ldots, x_R)$ be a point in V' above it. Since V is symmetric with respect to sign changes on the last $R - r$ coordinates, we can then make such sign changes on $x_{r+1}, \ldots, x_R$ to achieve the desired positivity condition. (I am indebted to Roger Howe for the above proof.)

We let j_0 be the index such that $x_j < x_{j_0}$ for all $j \neq j_0$. After replacing $x_1, \ldots, x_R$ by $c x_1, \ldots, c x_R$ for some real number $c > 0$, we may assume that

$$0 < x_j < p^{-m} \quad \text{for all } j = 1, \ldots, R.$$

We then apply Lemma 3.1, (i) and (ii) with $\beta_i = \eta_i$ for $i = 1, \ldots, r$. Let n and α be as in that lemma. If ε is small enough, we obtain

$$0 < \sum_{i=1}^{r} a_{ji} p^{-(n+1)} s_n(\alpha \eta_i) < p^{-m} \quad \text{for } j = r+1, \ldots, R.$$

35

Hence

$$0 < \sum a_{ji} s_n(\alpha\eta_i) < p^{n+1-m} < p^{n+1}.$$

Therefore

$$s_{n-m}(\alpha\eta_j) = s_n(\alpha\eta_j) = \sum a_{ji} s_n(\alpha\eta_i).$$

We take $\alpha_1 = \alpha$ to satisfy the first part of Lemma 1.1.

For Lemma 1.1 (ii) and (iii), we select the scaling factor $c > 0$ such that

$$0 < x_j < p^{-m} \quad \text{for } j \neq j_0 \quad \text{but} \quad p^{-m} < x_{j_0} < 2p^{-m}.$$

We select $\alpha = \alpha_2$ in Lemma 3.1. Then the desired conditions are satisfied for $j \neq j_0$ as before, but for j_0, Lemma 3.1 now shows

$$s_{n-m}(\alpha\eta_{j_0}) = s_n(\alpha\eta_{j_0}) - p^{n-m+1}.$$

This concludes the proof.

Measures in the Composite Case 12

In Chapter 4 we developed the formalism of associated power series for measures on $\mathbf{Z}_p$. It is necessary to develop it in general. We do this in the present chapter, which could (and should) have been done immediately following Chapter 4. In all this work, a prime p is given a special role. Values of functions lie in $\mathbf{C}_p$. In dealing with this composite case, it is also useful to follow Katz, and associate to a measure not only a power series, but an analytic function on the "formal multiplicative group." This is explained in §2. The introduction of additional notation to handle this composite case, however, made it worthwhile to separate the two cases. Measures on $\mathbf{Z}_p$ itself, without the extra d, occur both in their own right, and as auxiliaries to the composite case, so it is useful to have their properties tabulated separately.

The present chapter is independent of everything else in the book, and can be omitted by those who wish to read at once the results of Chapter 13, needed to bound the plus part of the class number in terms of the minus part, for the Ferrero–Washington theorems.

§1. Measures and Power Series in the Composite Case

In Chapter 4, we dealt only with the formalism of measures and power series on $\mathbf{Z}_p$. To handle characters with conductor dp^n where d is a positive integer prime to p, one has to deal with the composite case. Thus we now give an exposition of the formalism in this more general context.

Let Z be a profinite group, equal to the projective limit of its quotients

$$Z = \lim_{H} Z/H,$$

where H ranges over the open subgroups of finite index. Let $\mathfrak{o}$ be a complete valuation subring of the p-integers in $\mathbf{C}_p$. We let the **Iwasawa algebra** be

$$\Lambda_{\mathfrak{o}}(Z) = \lim_{H} \mathfrak{o}\,[Z/H],$$

where $\mathfrak{o}[Z/H]$ is the group ring of Z/H over $\mathfrak{o}$. We recall that an $\mathfrak{o}$-valued measure on the projective system $\{Z/H\}$ is a family of $\mathfrak{o}$-valued functions $\{\mu_H\}$, which is a distribution. This means: given $H' \subset H$, we have

$$\mu_H(x) = \sum_y \mu_{H'}(y),$$

where $x \in Z/H$ and the sum is taken over $y \in Z/H'$ lying above x under the canonical map $Z/H' \to Z/H$. The association

$$\mu_H \mapsto \sum_{x \in Z/H} \mu_H(x)x \in \mathfrak{o}[Z/H]$$

lifts to an isomorphism between the additive group of $\mathfrak{o}$-valued measures and the Iwasawa algebra.

Observe that the product in $\Lambda_{\mathfrak{o}}(Z)$ corresponds to the convolution of measures.

If φ is an $\mathfrak{o}$-valued function on Z, factoring through some factor group Z/H, in other words, if φ is a step function, then we define its **integral**

$$\int \varphi\, d\mu = \sum_{x \in Z/H} \varphi(x)\mu_H(x).$$

This value is independent of the choice of H, by the distribution relation. The integral extends to the space of $\mathfrak{o}$-valued continuous functions on Z by uniform approximation with step functions.

Let $C(Z, \mathfrak{o})$ be the $\mathfrak{o}$-space of continuous functions of Z into $\mathfrak{o}$, with sup norm. There is a bijection between $\mathfrak{o}$-valued measures on Z and $\mathfrak{o}$-valued $\mathfrak{o}$-linear functionals

$$\lambda: C(Z, \mathfrak{o}) \to \mathfrak{o}.$$

Indeed, the measure μ gives rise to the functional

$$\varphi \mapsto \int \varphi\, d\mu.$$

Conversely, given λ, if $x \in Z/H$ and φ_x is the characteristic function of the set of all $y \in Z$ such that $y \equiv x \bmod H$, then we define

$$\mu_H(x) = \lambda(\varphi_x).$$

It is easily verified that these associations are inverse to each other, and establish a norm-preserving bijection, where

$$\|\mu\| = \sup_{x, H} |\mu_H(x)|$$

is the sup norm.

Now suppose that there is a finite subgroup Z_0 such that

$$Z = Z_0 \times \mathbf{Z}_p.$$

Let Γ be a multiplicative group isomorphic to $\mathbf{Z}_p$, with topological generator γ, so the isomorphism is given by

$$z \mapsto \gamma^z, \quad \text{with } z \in \mathbf{Z}_p.$$

Let $H_n = \{1\} \times p^n \mathbf{Z}_p$, and let $\gamma_n = \gamma \mod \Gamma^{p^n}$. Then

$$Z_n = Z/H_n \cong Z_0 \times \Gamma_n,$$

where $\Gamma_n = \Gamma/\Gamma^{p^n}$ is cyclic, with generator γ_n.

Let X be a variable, $T = 1 + X$, and

$$h_n(X) = (1 + X)^{p^n} - 1.$$

Then the element of $\mathfrak{o}[Z_n]$ corresponding to μ_{H_n} is a linear combination

(1)
$$P_n(X) = \sum_{\sigma \in Z_0} \sum_{r=0}^{p^n - 1} \mu_{n, \sigma}(r) \sigma \gamma_n^r$$

$$\equiv \sum_{\sigma \in Z_0} \sum_{r=0}^{p^n - 1} \sum_{k=0}^{r} \mu_{n, \sigma}(r) \binom{r}{k} \sigma X^k \mod h_n(X).$$

where $\gamma_n = T \mod h_n$. For each σ, the map

$$r \mapsto \mu_{n, \sigma}(r)$$

determines a measure μ_σ on $\mathbf{Z}_p$ as discussed in Chapter 4. This makes it possible to extend results concerning measures on $\mathbf{Z}_p$ to measures on the more general groups now being considered. If we let

$$P_n(X) = \sum_{k=0}^{p^n - 1} c_{n, k} X^k,$$

then the coefficients

$$(2) \qquad c_{n,k} = \sum_{\sigma} \sum_{r} \mu_{n,\sigma}(r) \binom{r}{k} \sigma = \sum_{\sigma} c_{n,k}(\sigma) \sigma$$

lie in the group ring $\mathfrak{o}[Z_0]$.

We let $P(X)$ be the projective limit of the polynomials $P_n(X)$. Then $P(X)$ is an element of $\mathfrak{o}[Z_0][[X]]$, and we shall write the correspondence between μ and the power series P by

$$f = P\mu, \qquad \mu = \mu_f.$$

Let us write

$$P(X) = \sum_{k=0}^{\infty} c_k X^k = \sum_{k=0}^{\infty} \sum_{\sigma \in Z_0} c_{k,\sigma} \sigma X^k.$$

Then the coefficients $c_{k,\sigma}$ are given by the integrals

$$(3) \qquad c_{k,\sigma} = \int_{Z_p} \binom{x}{k} d\mu_\sigma(x),$$

because we can apply Theorem 1.1 of Chapter 4.

Let φ be a continuous function on Z. Then for each σ, we get a function

$$\varphi_\sigma(r) = \varphi(\sigma, r) \quad \text{with } r \in Z_p.$$

If φ factors through Z_n, then

$$(4) \qquad \int_Z \varphi \, d\mu = \sum_{\sigma} \sum_{r \in Z(p^n)} \varphi(\sigma, r)\mu_n(\sigma, r)$$

$$= \sum_{\sigma \in Z_0} \int_{Z_p} \varphi_\sigma \, d\mu_\sigma.$$

Any continuous function φ on Z can be viewed as a family of continuous functions $\{\varphi_\sigma\}$ on Z_p. Hence

$$(5) \qquad \int_Z \varphi \, d\mu = \sum_{\sigma \in Z_0} \int_{Z_p} \varphi_\sigma \, d\mu_\sigma.$$

Using Mahler's Theorem 1.3 of Chapter 4, if we write

$$\varphi_\sigma(x) = \sum a_{n,\sigma} \binom{x}{n}$$

and the power series associated with μ_σ is

$$(6) \qquad\qquad f_\sigma(X) = \sum c_{n,\sigma} X^n,$$

then

$$(7) \qquad\qquad \int_Z \varphi \, d\mu = \sum_\sigma \sum_n c_{n,\sigma} a_{n,\sigma}.$$

The sum formula (5) in principle allows us to reduce the study of any measure μ_f to the individual measures associated with the power series f_σ on the fiber $\{\sigma\} \times \mathbf{Z}_p$. The formulas of Chapter 4 apply to each f_σ. In addition, we have trivial ones relating to the extra factor Z_0. For instance:

(8)   *If φ is a function on Z_0, then*

$$\varphi \mu_f = \mu_g \quad \text{where } g_\sigma = \varphi(\sigma) f_\sigma.$$

This applies in particular to the characteristic function φ_α of a single element $\alpha \in Z_0$. Then

$$\varphi_\alpha \mu_f = \mu_{f_\alpha}$$

where f_α is identified with the element $f_\alpha \cdot \alpha$ in $\mathbf{Z}_p[Z_0][[X]]$. In this last example, we have of course the Fourier expansion of φ_α given by

$$\varphi_\alpha = \frac{1}{|Z_0|} \sum_\psi \bar{\psi}(\alpha) \psi,$$

where the sum is taken over all characters ψ of Z_0.

At any finite level, that is on any one of the groups

$$Z_0 \times \mathbf{Z}(p^n),$$

the space of functions is generated by the product functions

$$\varphi_0 \otimes \varphi_p,$$

where φ_0 is a function on Z_0 and φ_p is a function on $\mathbf{Z}(p^n)$. By definition,

$$(\varphi_0 \otimes \varphi_p)(\sigma, x_p) = \varphi_0(\sigma) \varphi_p(x_p).$$

41

Thus to test whether two measures are equal, it suffices to test whether their integrals on such functions are equal. These integrals decompose as simple sums according to (4) and (5), namely

$$(9) \qquad \int_Z \varphi_0(\sigma)\varphi_p(x_p)\, d\mu(\sigma, x_p) = \sum_\sigma \varphi_0(\sigma) \int_{Z_p} \varphi_p(x_p)\, d\mu_\sigma(x_p).$$

This applies for instance if φ is a character of Z which factors through a finite level.

Let f be the power series associated with μ. Given the function φ_0 on Z_0, the expression of (9) defines a measure μ_{φ_0} on Z_p whose associated power series is

$$f_{\varphi_0} = \sum_\sigma \varphi_0(\sigma) f_\sigma, \quad \text{while} \quad \mu_{\varphi_0} = \sum_\sigma \varphi_0(\sigma)\mu_\sigma.$$

In other words, we have the formula

$$(10) \qquad \int_Z \varphi_0 \otimes \varphi_p\, d\mu_f = \int_{Z_p} \varphi_p\, d\mu_g \quad \text{where } g = f_{\varphi_0}.$$

If $Z_0 = \mathbf{Z}(d) = \mathbf{Z}/d\mathbf{Z}$, then we may take for φ_0 an additive character, which is of the form

$$\psi_0 : x_0 \mapsto \zeta_0^{x_0}, \qquad x_0 \in \mathbf{Z}(d),$$

for some d-th root of unity ζ_0. In that case, the measure μ_g above will also be denoted by

$$\mu_{\zeta_0} \quad \text{or} \quad \mu_{\psi_0}$$

if ψ_0 is the above character.

Let $N = dp^n$. An N-th root of unity ζ has a unique product decomposition

$$\zeta = \zeta_0 \zeta_p$$

where ζ_0 is a d-th root of unity, and ζ_p is a p^n-th root of unity. Then we have, by definition and (10):

$$(11) \qquad \int_Z \zeta_0^{x_0}\zeta_p^{x_p}\varphi_p(x_p)\, d\mu(x) = \int_{Z_p} \zeta_p^{x_p}\varphi_p(x_p)\, d\mu_{\zeta_0}(x_p).$$

In particular, for $\varphi_p = 1$,

(12)
$$\int_Z \zeta_0^{x_0} \zeta_p^{x_p} \, d\mu(x) = \int_{\mathbf{Z}_p} \zeta_p^{x_p} \, d\mu_{\zeta_0}(x_p)$$

where μ_{ζ_0} is the measure whose associated power series is

(13)
$$f_{\zeta_0} = \sum_{x_0 \in \mathbf{Z}(d)} \zeta_0^{x_0} f_{x_0}.$$

For this formalism it is therefore useful to define $x \in Z$ by its two components, $x = (x_0, x_p)$, where $x_0 \equiv x \bmod d$, and $x_p \equiv x \bmod p^n$. Then

$$\zeta^x = \zeta_0^{x_0} \zeta_p^{x_p},$$

so that the above integral can be written more simply with ζ^x.

By the orthogonality of characters, it is then immediate that we can recover f_{x_0} from f_{ζ_0} by means of the formula

(14)
$$f_{x_0} = \frac{1}{d} \sum_{\zeta_0} \zeta_0^{-x_0} f_{\zeta_0}.$$

§2. The Associated Analytic Function on the Formal Multiplicative Group

Katz [Ka 3] has shown that in addition to the power series associated with a measure, it is also useful to associate an analytic function on the "formal multiplicative group." The formalism is similar to the one of power series developed in Chapter 4, but it is more convenient in some situations, especially when dealing with the extra factor $\mathbf{Z}(d)$ in the composite case. Again, Katz's formalism makes certain constructions due to Iwasawa and Leopoldt appear completely natural, and we reproduce this formalism below.

We fix a positive integer d prime to p. In the sequel we let N denote any positive integer of the form dp^n, and again we let $Z = \mathbf{Z}(d) \times Z_p$, so that $Z_0 = \mathbf{Z}(d)$. Let

$$T = 1 + X$$

as usual. Let $\mathfrak{m}$ be the maximal ideal in $\mathfrak{o}_{\mathbf{C}_p}$. A function R on $1 + \mathfrak{m}$ is called **analytic** if there exists a power series f with coefficients in $\mathbf{C}_p$, converging on $\mathfrak{m}$, such that

$$R(1 + z) = f(z) \quad \text{for all } z \in \mathfrak{m}.$$

Let η be a d-th root of unity. We say that R is **analytic** on $\eta(1 + \mathfrak{m})$ if there exists a power series f as above such that

$$R(\eta(1 + z)) = f(z) \quad \text{for all } z \in \mathfrak{m}.$$

Then f is uniquely determined by R, and is said to be **associated** with R.

Example. The function $\log T = \log(1 + X)$ is analytic in the above sense. In the applications, we shall deal principally with this log, or with rational functions.

We let $\mathbf{G}_p(d)$ be the (disjoint) union

$$\mathbf{G}_p(d) = \bigcup_{\eta^d = 1} \eta(1 + \mathfrak{m}).$$

A function R is called **analytic** on $\mathbf{G}_p(d)$ if it is analytic on each "component" $\eta(1+\mathfrak{m})$. Then R consists of a family $\{R_\eta\}$ where each R_η is analytic on the corresponding component. If $\zeta \in \eta(1+\mathfrak{m})$, we use the notation

$$\zeta = \eta u, \qquad \eta = \zeta_0 = \omega(\zeta), \qquad u = \zeta_p = \langle \zeta \rangle_p.$$

Then

$$R(\zeta) = R_{\omega(\zeta)}(\zeta) = f_{\omega(\zeta)}(\zeta_p - 1).$$

We shall call $\mathbf{G}_p(d)$ the **formal multiplicative group** (at p, of level d).

Remark. The group $\mathbf{G}_p(d)$ may be viewed as the group of (continuous) $\mathbf{C}_p^*$-valued characters on $Z = \mathbf{Z}(d) \times \mathbf{Z}_p$, by the mapping

$$\psi_\zeta : x \mapsto \zeta^x = \zeta_0^{x_0} \zeta_p^{x_p},$$

for each $\zeta \in \mathbf{G}_p(d)$.

Let μ be a measure on Z, with values in $\mathfrak{o}$. An analytic function R on $\mathbf{G}_p(d)$ will be said to be **associated with** μ if the power series associated with R have coefficients in $\mathfrak{o}$, and if we have the formula

$$\int_Z \zeta^x \, d\mu(x) = R(\zeta)$$

for all N-th roots of unity ζ, $N = dp^n$ (all n). The Weierstrass preparation theorem shows that an associated analytic function R, if it exists, is uniquely determined by μ. Of course, we want the above formula to be true also for all $\zeta \in \mathbf{G}_p(d)$, but we can prove it later, when ζ is not a root of unity.

We shall say that μ is **rational** if R is a rational function of T (and hence of X). Thus the right-hand side $R(\zeta)$ is just the value of this rational function at $T = \zeta$. We then call R the **associated rational function**. We also observe that the rational function is the same (if it exists) as that associated with the restriction of μ to

$$\{0\} \times \mathbf{Z}_p.$$

Indeed, we have from the definition of a distribution:

$$R(\zeta_p) = \int_Z \zeta_p^{x_p} \, d\mu(x) = \int_{\mathbf{Z}_p} \zeta_p^{x_p} \, d\mu(0, x_p).$$

R 1. *A measure always has an associated analytic function R. If $d = 1$ and μ is a measure on $\mathbf{Z}_p$, then*

$$R(T) = f(X)$$

is the associated power series of Chapter 4.

Proof. For a measure on $\mathbf{Z}_p$ with $d = 1$, this follows from **Meas 0** and **Meas 2** of Chapter 4, §2. The general case then follows at once. Note that in **Meas 0**, we have

$$f(0) = R(1).$$

Furthermore, if μ is a measure on $Z = \mathbf{Z}(d) \times \mathbf{Z}_p$, then the measure μ_{ζ_0} at the end of the last section is a measure on $\mathbf{Z}_p$ to which we can apply **R 1**, with $d = 1$.

R 2. *If a measure μ on Z has an associated rational function $R(T)$, then the measure μ_{ζ_0} has an associated rational function, given by*

$$R_{\zeta_0}(T) = R(\zeta_0 T) = f_{\zeta_0}(X).$$

Proof. This is immediate from the definition of the associated rational function, and the integral formula (12) of §1.

Lemma 2.1. *Assume that μ is a measure on Z such that each μ_{ζ_0} is rational, for every d-th root of unity ζ_0. Assume also that the functions*

$$R_{\zeta_0}(\zeta_0^{-1} T) = R(T)$$

are independent of ζ_0. Then μ is rational, and its associated rational function is $R(T)$.

Proof. This is again immediate from the integral formula (12) of §1.

R 3. *Let μ be a measure having associated analytic function R. Then the formula*

$$\int_Z \zeta^x \, d\mu(x) = R(\zeta)$$

holds for any $\zeta \in G_p(d)$.

Proof. This is a direct consequence of **R 1**, formula (12) of §1, and the analogous result for measures on Z_p stated in Theorem 1.2 of Chapter 4.

R 4. *Let $\zeta \in G_p(d)$. Let R be the associated analytic function of μ. Then*

$$\zeta^x \mu(x) \text{ has associated analytic function } R(\zeta T).$$

Proof. Immediately from **R 3**, since $\zeta_1^x \zeta_2^x = (\zeta_1 \zeta_2)^x$.

Just as with associated power series, **R 4** allows us to use the Fourier expansion of any step function to obtain its associated analytic function. Thus let φ be a step function of level M which is a divisor of dp^n (possibly a pure power of p). Then

$$\varphi(x) = \sum_{\zeta^M = 1} \hat{\varphi}(\zeta) \zeta^x$$

and

$$\hat{\varphi}(\zeta) = \frac{1}{M} \sum_{x \in Z(M)} \varphi(x) \zeta^{-x}.$$

We shall now give a first example of the use of such expansions in connection with the **unitization operator** U defined by the formula

$$UR(T) = R(T) - \frac{1}{p} \sum_{\zeta^p = 1} R(\zeta T).$$

Let

$$Z^* = Z(d) \times Z_p^*.$$

R 5. *Let φ be the characteristic function of Z^*. If R is associated with μ, then*

$$UR \text{ is associated with } \varphi\mu.$$

Proof. The Fourier expansion of the characteristic function of Z_p^* was already computed in Chapter 4, §2 and is trivially determined to be given by

$$\hat{\varphi}(\zeta) = \begin{cases} -1/p & \text{if } \zeta \neq 1. \\ \dfrac{p-1}{p} & \text{if } \zeta = 1, \end{cases}$$

and ζ ranges over p-th roots of unity. Property **R 5** then follows from **R 4** and the Fourier expansion.

R 6. *Let $N = dp^n$ where $n \geq 0$. Let χ be a Dirichlet character of conductor N, and let ζ be a primitive N-th root of unity. If R is the analytic function associated with μ, then the analytic function associated with $\chi\mu$ is*

$$\frac{S(\chi, \zeta)}{N} \sum_{a \in Z(N)^*} \bar{\chi}(a) R(\zeta^{-a} T).$$

where

$$S(\chi, \zeta) = \sum \chi(a) \zeta^a.$$

Proof. The computation of the Fourier transform of χ is routine and is left to the reader. (One has of course to use Theorem 1.1 of Chapter 3, §1.)

R 7. *Let R be the analytic function associated with μ. Let $D = TD_T$ and let k be an integer ≥ 0. Then*

$$x_p^k \mu(x) \text{ has associated analytic function } D^k R(T).$$

The same statement holds if μ is rational, and "analytic" is replaced by "rational" in the above statement.

Proof. We may use $\varphi_p(x_p) = x_p^k$ in (11) of §1, and then apply **Meas 6** of Chapter 4, §2 to each one of the measures μ_{ζ_0}, after using **R 1**. Thus the general case is reduced to the special case of measures on Z_p.

R 8. *Let* $R = UR$ *be the associated analytic function of a measure* μ *with support in* Z^*. *Then the measure*

$$x_p^{-1}\mu(x) \text{ has associated analytic function } \mathbf{U}H,$$

where H *is any analytic function on* $\mathbf{G}_p(d)$ *such that*

$$DH = R.$$

Proof. Since $x_p^{-1}\mu(x)$ is a measure on Z^*, there exists an analytic function F associated with it. Then **R 7** yields

$$DF = R.$$

We can let $H = F$. Since the kernel of D consists of the functions which are constant on each component of $\mathbf{G}_p(d)$, if we select any H such that $DH = R$, then $\mathbf{U}H$ will have the same value, independent of the choice of H, thus proving the desired property.

The "integration" of the analytic function $R(T)$ can be performed formally in terms of the variable X. Indeed, say on the coset $1 + \mathfrak{m}$, if R is defined by a power series $f(X)$ then an analytic function H such that $DH = R$ is given in terms of X by

$$H(T) = \int F(X) \frac{dX}{1+X} = \int \frac{F(X)}{1+X} dX = \int R(T) \frac{dT}{T} .$$

Remark. The formalism of this section is set up to extend at once to its adelization over d prime to p.

§3. Computation of $L_p(1, \chi)$ in the Composite Case

Let $E_{1,c}$ be the measure defined in Chapter 10, §2, giving rise to the p-adic L-function and the Bernoulli distribution, regularized with c so as to be integral valued. We shall apply the previous considerations to this measure.

By definition, we have for $s = 0$ the value

$$L_p(1, \chi) = \frac{-1}{1 - \chi(c)} \int_Z \chi(a) a_p^{-1} \, dE_{1,c}(a).$$

The conductor of χ being $dp^n = N$, the integral might as well be taken over

$$Z^{**} = Z(d)^* \times Z_p^*.$$

In any case, if we find (as we shall do below) that $E_{1,c}$ has an associated rational function, then the general formalism also yields successively the corresponding analytic function for the measure

$$\chi(a)a_p^{-1}E_{1,c}(a).$$

We may then evaluate this analytic function at $T = 1$ to get the value $L_p(1, \chi)$.

We shall now carry out on $\mathbf{Z}(d) \times \mathbf{Z}_p$ the same analysis that we did for $\mathbf{Z}_p$ in Chapter 4, concerning the associated power series, and the Leopoldt formula for the value of the L-function at $s = 1$.

We recall that if c is a positive integer prime to dp, then the measure $E_{1,c}$ on $\mathbf{Z}_p$ has an associated rational function (equal to its associated power series by **R 1**), which is

$$R_{1,c}(T) = \frac{1}{T-1} - \frac{c}{T^c - 1}$$

according to Proposition 3.4 of Chapter 4.

Proposition 3.1. *Let c be a positive integer prime to dp. Then $E_{1,c}$ on $\mathbf{Z}(d) \times \mathbf{Z}_p$ has an associated rational function, equal to $R_{1,c}(T)$ above.*

Proof. To extend the above result from $\mathbf{Z}_p$ to $\mathbf{Z}(d) \times \mathbf{Z}_p$, it suffices to prove that for every root of unity $\zeta \in \mathbf{\mu}_N$, and $N = dp^n$, we have

$$\int_Z \zeta^x \, dE_{1,c}(x) = R_{1,c}(\zeta).$$

By definition, essentially (cf. formula **B 6** of Chapter 2, §2) we know that for any function φ on $\mathbf{Z}(N)$ we have

$$\sum_{k=0}^{\infty} B_{k,\varphi} \frac{Z^k}{k!} = \sum_{a=0}^{N-1} \varphi(a) \frac{Ze^{aZ}}{e^{NZ} - 1}.$$

We apply this to $\varphi(x) = \zeta^x$ and $\varphi(x) = \zeta^{cx}$. Summing a geometric series from 0 to $N-1$ then yields

$$\sum_{k=1}^{\infty} \left(\frac{1}{k} B_{k,\varphi} - c^k \frac{1}{k} B_{k,\varphi \circ c} \right) \frac{Z^{k-1}}{(k-1)!} = \frac{1}{\zeta T - 1} - \frac{c}{\zeta^c T^c - 1},$$

where $T = e^Z$. The right-hand side at $T = 1$ is the same as $R_{1,c}(\zeta)$, and is also the same as the left-hand side at $Z = 0$. But that is precisely the value of the desired integral for $k = 1$. This proves Proposition 3.1.

Proposition 3.2. *Let* χ *have conductor* $N = dp^n$ *with* $n \geq 0$. *Then* $\chi E_{1,c}$ *has an associated rational function* $R_{\chi,c}$ *given by the formula*

$$R_{\chi,c}(T) = G_\chi(T) - c\chi(c)G_\chi(T^c),$$

where for any primitive N-*th root of unity* ζ,

$$G_\chi(T) = \frac{S(\chi, \zeta)}{N} \sum_{a \in Z(N)^*} \bar\chi(a) \frac{1}{\zeta^{-a}T - 1}.$$

Proof. Special case of **R 6**.

We can write $R_{\chi,c}(T)$ in full in the form:

$$R_{\chi,c}(T) = \frac{S(\chi, \zeta)}{N} \sum_a \bar\chi(a) \left[\frac{\zeta^a}{T - \zeta^a} - \frac{c\chi(c)\zeta^a}{T^c - \zeta^a} \right].$$

This puts us in the position of applying **R 8**, and of finding an analytic function $H_{\chi,c}$ such that $H_{\chi,c}(1) = 0$, and

$$DH_{\chi,c} = R_{\chi,c}.$$

We let λ range over c-th roots of unity, and we let

$$H_{\chi,c}(T) = \frac{-S(\chi, \zeta)}{N} \sum_{\lambda \neq 1} \sum_{a \in Z(N)^*} \bar\chi(a) \log \frac{T - \lambda\zeta^a}{1 - \lambda\zeta^a}.$$

Exactly the same verification as in Chapter 4, §3 (or a direct integration using partial fractions) shows that this value for $H_{\chi,c}(T)$ satisfies our requirements.

By **R 6**, the analytic function associated with the measure $\chi(a)a_p^{-1}E_{1,c}(a)$ is $UH_{\chi,c}(T)$.

Theorem 3.3. *Let* χ *have conductor* $N = dp^n$, $n \geq 0$. *Let* ζ *be a primitive* N-*th root of unity. Then*

$$L_p(1, \chi) = - \left(1 - \frac{\chi(p)}{p} \right) \frac{S(\chi, \zeta)}{N} \sum_{a \in Z(N)^*} \bar\chi(a) \log_p(1 - \zeta^a).$$

Proof. Let ξ range over the p-th roots of unity. By **R 6**,

$$-(1 - \chi(c))L_p(1, \chi) = UH_{\chi,c}(1).$$

Hence, following exactly the proof of Chapter 4, Theorem 3.6:

$$-(1 - \chi(c))L_p(1, \chi) = \frac{1}{p} \frac{S(\chi, \zeta)}{N} \sum_\xi \sum_a \sum_{\lambda \neq 1} \bar{\chi}(a) \log_p\left(\frac{\xi - \lambda\zeta^a}{1 - \lambda\zeta^a}\right)$$

$$= \frac{1}{p} \frac{S(\chi, \zeta)}{N} \sum_a \bar{\chi}(a) \log_p\left(\prod_{\lambda \neq 1} \prod_\xi \frac{\xi - \lambda\zeta^a}{1 - \lambda\zeta^a}\right).$$

But

$$\prod_{\lambda \neq 1} \prod_\xi \frac{\xi - \lambda\zeta^a}{1 - \lambda\zeta^a} = \prod_{\lambda \neq 1} \frac{1 - \lambda\xi^{ap}}{(1 - \lambda\zeta^a)^p}.$$

If $p \mid N$, then as in Theorem 3.6 of Chapter 4 we find that

$$\sum_a \bar{\chi}(a) \log_p(1 - \lambda\zeta^{ap}) = 0,$$

and the rest of the proof is identical with the previous one. If $p \nmid N$, then we change variables, letting $a \mapsto p^{-1}a \bmod N$. We then find

$$\sum_a \bar{\chi}(a) \log_p(1 - \lambda\zeta^{ap}) = \chi(p) \sum_a \bar{\chi}(a) \log_p(1 - \lambda\zeta^a).$$

Thus we get an extra term besides that of Chapter 4, which gives rise to the factor $(1 - \chi(p)/p)$ as stated in the theorem. Except for that, the proof is again identical with the previous one.

13 Divisibility of Ideal Class Numbers

Classical results of Kummer give bounds for certain class numbers and estimate, for instance, the plus part in terms of the minus part for cyclotomic fields.

Such estimates have been carried out more systematically and generally by Iwasawa, to include estimates for his invariants, and will be given in the first two sections of this chapter.

Next we deal with the l-part of the class number in a cyclic extension of degree p^n when p is a prime $\neq l$. We also give examples of Iwasawa of non-cyclotomic $\mathbf{Z}_p$-extensions when the order of the class number grows exponentially. These examples are based on a classical formula (Takagi–Chevalley) expressing the fact that in a highly ramified extension, the ramified primes have a strong tendency to generate independent ideal classes.

We conclude with a lemma of Kummer which is still somewhat isolated. In this connection, cf. [Wa 5].

§1. Iwasawa Invariants in $\mathbf{Z}_p$-extensions

Let K be a number field and K_∞ a $\mathbf{Z}_p$-extension. We let K_n denote the subfield of degree p^n over K, so $K_0 = K$.

Let C_n be the p-primary part of the ideal class group of K_n. We also write $C(K)$ for the *p-primary part* of the ideal class group of K. Then by Iwasawa theory of Chapter 5, we have

$$|C_n| = p^{e_n} \quad \text{where } e_n = mp^n + \lambda n + O(1).$$

We call m, λ the **Iwasawa invariants** of the $\mathbf{Z}_p$-extension. We call m the **exponential invariant**, and λ the **linear invariant**. We indicate the dependence of m and λ on the $\mathbf{Z}_p$-extension by the notation.

$$m = m(K_\infty/K) \quad \text{and} \quad \lambda = \lambda(K_\infty/K).$$

If K_∞ is the cyclotomic $\mathbf{Z}_p$-extension, then these invariants depend only on K. We may then write more briefly

$$m = m(K) \quad \text{and} \quad \lambda = \lambda(K).$$

Put

$$V_K = C(K_\infty/K) = \varprojlim C(K_n).$$

From the structure theorem of Chapter 5, we have a quasi-isomorphism

$$\boxed{V_K \sim \prod \Lambda/p^{m_i} \oplus \prod \Lambda/f_j.}$$

For any abelian group V we let $V^{(p)}$ be its p-primary part, i.e. the subgroup of elements annihilated by a power of p. Then we have quasi-isomorphisms

$$V_K^{(p)} \sim \prod \Lambda/p^{m_i} \quad \text{and} \quad V_K/V_K^{(p)} \sim \prod \Lambda/f_j.$$

From the structure Theorems 1.2 and 1.3 of Chapter 5, we know that

$$\boxed{\lambda(K_\infty/K) = \sum \deg f_j.}$$

Furthermore, we have the characterization:

λ *is the rank of* $V_K/V_K^{(p)}$ *as a finitely generated* $\mathbf{Z}_p$-*module.*

For the exponential invariant, we have

$$\boxed{m = \sum m_i.}$$

We shall now compare these invariants in a $\mathbf{Z}_p$-extension and its lifting over a finite extension.

First note that if $F \subset K$ and F_∞ is a $\mathbf{Z}_p$-extension of F, then $K_\infty = KF_\infty$ is a $\mathbf{Z}_p$-extension of K, and there is some integer r such that

$$K_n = KF_{n+r}.$$

Lemma 1. *Let $F \subset K$ be number fields. Let F_∞ be a $\mathbf{Z}_p$-extension of F. Then*

$$m(F_\infty/F) \leq m(KF_\infty/K) \quad and \quad \lambda(F_\infty/F) \leq \lambda(KF_\infty/K).$$

Proof. The degree $[K_n : KF_n]$ is bounded by a fixed power p^r. By class field theory, we have

$$(C(F_n) : N_{K_n/F_n} C(K_n)) \leq [K : F]p^r.$$

Hence

$$|C(K_n)| \gg |C(F_n)|,$$

and therefore

$$\operatorname{ord}_p C(K_n) \geq \operatorname{ord}_p C(F_n) - O(1).$$

This proves the first inequality in light of the formula for the orders.

For the second, let us use the functorial notation

$$C(F_\infty/F) = \varprojlim C(F_n).$$

The norm maps

$$N_{K_n/F_n} : C(K_n) \to C(F_n)$$

are compatible with the norm maps in the projective limits in the tower over F_0 and K_0 respectively, and thus induce a homomorphism

$$N : C(K_\infty/K) \to C(F_\infty/F).$$

Since the index of the image of N_{K_n/F_n} is bounded for all n (as in the first part of the proof), it follows that the image of N in $C(F_\infty/F)$ is of finite index, in other words that N is quasi-surjective.

Now put $V_F = C(F_\infty/F)$ and similarly for V_K. Let $V^{(p)}$ as usual denote the subgroup of elements annihilated by a power of p. Then we have a quasi-surjective homomorphism

$$V_K/V_K^{(p)} \to V_F/V_F^{(p)}.$$

The inequality for the λ-invariants then follows at once since λ is the rank of the above modules as finitely generated $\mathbf{Z}_p$-modules.

We have been concerned with the p-divisibility of the order of the ideal class group in $\mathbf{Z}_p$-extensions. An analogue of Lemma 1 can be given for any prime number.

Lemma 2. *Let F_∞ be a $\mathbf{Z}_p$-extension of a number field F. Let K be a finite extension of F and let $K_\infty = KF_\infty$. Let l be any prime number. If $\mathrm{ord}_l |C(K_n)|$ is bounded, then $\mathrm{ord}_l |C(F_n)|$ is bounded.*

Proof. The argument using the norm index in the first part of the proof of Lemma 1.1 applies equally well to prove the result stated in Lemma 1.2.

Next we study another Iwasawa invariant, the rank.
If A is an abelian group, we define the p-**rank**,

$$\mathbf{rank}_p(A) = \text{dimension of } A/pA \text{ over the prime field } \mathbf{F}_p.$$

We have a criterion for the vanishing of the exponential Iwasawa invariant in terms of this p-rank of the groups C_n. We phrase the criterion to apply in general to torsion modules over the Iwasawa algebra.

We use the notation of Chapter 5, Theorem 1.3, where we dealt with a module of Iwasawa type, and say

$$V \sim \prod_{i=1}^{r} \Lambda/p^{m_i} \oplus \prod_{j=1}^{t} \Lambda/f_j,$$

where the f_j are distinguished polynomials. We let

$$r_1(V) = r$$

be the number of factors of type Λ/p^m for some m. As in Chapter 5, we let $v_1, \ldots, v_s$ be elements of V such that if we put

$$U_0 = \mathbf{Z}_p\text{-submodule of } V \text{ generated by } (\gamma - 1)V \text{ and } v_1, \ldots, v_s,$$
$$U_n = g_n U_0 \text{ with } g_n = 1 + \gamma + \cdots + \gamma^{p^n - 1},$$

then

$$V_n = V/U_n$$

is finite for all n. We put $e_n = e(V_n) = \mathrm{ord}_p V_n$. The proofs of Theorem 1.2 and Theorem 1.3 of Chapter 5 distinguish the two cases of modules of type

$$\Lambda/p^m \quad \text{and} \quad \Lambda/f$$

where f is distinguished. They immediately show the following result.

Lemma 3. *We have* $\operatorname{rank}_p V_n = r_1(V)p^n + O(1)$, *where* $r_1(V)$ *is the number of factors of type* Λ/p^m *above. Furthermore, the following conditions are equivalent.*

(i) *All the factors* Λ/p^{m_i} *are equal to 0, so* $r_1(V) = 0$ *and*

$$V \sim \prod \Lambda/(f_j).$$

(ii) *The p-rank of* V_n *is bounded independently of n.*

(iii) *In the order formula* $e_n = mp^n + \lambda n + O(1)$, *the exponential invariant m is equal to 0.*

§2. CM Fields, Real Subfields, and Rank Inequalities

We now wish to make a comparison of the behavior of the ideal class group in a field and a real subfield. We prove Kummer's theorem (Theorem 2.2) and Kummer type theorems in the following context.

A number field K is said to be a CM **field** (complex multiplication field) if it is a totally imaginary quadratic extension of a totally real field. We leave it as an exercise to prove:

K is a CM field if and only if the following condition is satisfied. Let ρ *be complex conjugation. Then* $\rho\sigma = \sigma\rho$ *for all embeddings* σ *of K into the complex numbers, and K is not real.*

The totally real subfield of K is then uniquely determined, and denoted by K^+. It also follows that a composite of CM fields is a CM field.

Although we are primarily interested in the cases when the CM field is abelian over the rationals, it is just as easy to deal with the more general case. However, we have to restate some results of Chapter 3 in this context.

Let K be a CM field. Let W_K be the group of roots of unity in K. The **Hasse index** Q_K can be defined as for abelian fields over the rationals, namely

$$Q_K = (E_K : W_K E_K^+),$$

where E_K is the group of units in K, while $E_K^+ = E(K^+)$ is the group of units in the real subfield.

We have:

Lemma 1. $Q_K = 1$ *or 2.*

The proof given for Theorem 4.1 of Chapter 3 applies here. In fact one verifies immediately that the map $u \mapsto \bar{u}/u$ gives an injection

$$E_K/W_K E_K^+ \rightarrow W_K/W_K^2.$$

Lemma 2. *Let K be a CM field and p a prime. Let $C^{(p)}(K)$ be the p-primary part of the ideal class group of K.*

(i) *The natural map $C^{(p)}(K^+) \to C^{(p)}(K)$ is injective if p is odd, and its kernel has order 1 or 2 if $p = 2$.*

(ii) *The norm map*

$$N_{K/K^+} : C(K) \to C(K^+)$$

is surjective.

Proof. For p odd this is a special case of the similar (trivial) lemma concerning ideal classes in extensions whose degree is prime to p. The proof will be given in full in the next section when we deal with this case for its own sake.

Suppose $p = 2$. Let $\mathfrak{a}$ be an ideal of K^+ and suppose $\mathfrak{a} = (\alpha)$ with α in K. Then $\bar{\alpha}/\alpha$ is a unit, and in fact a root of unity (the absolute value of all its conjugates is 1). We had defined above the map $\varphi : E_K \to W_K$ by $u \mapsto \bar{u}/u$. The association $\mathfrak{a} \mapsto \bar{\alpha}/\alpha$ then gives a well defined map

$$\operatorname{Ker}[C(K^+) \to C(K)] \to W_K/\varphi(E_K),$$

which is immediately verified to be injective. Hence the kernel has order 1 or 2. This proves (i).

The proof of (ii) is identical with that given for Theorem 4.3 of Chapter 3. It is a routine lemma of class field theory.

The eigenspace $C(K)^-$ is the kernel of the norm map N_{K/K^+} in $C(K)$, and for an odd prime p, we can identify the p-primary parts

$$C^{(p)}(K^+) = C^{(p)}(K)^+.$$

Unless otherwise specified, we continue to assume that p is an odd prime.

Let $K = K_0$ be a CM field, and let K_∞ be a $\mathbf{Z}_p$-extension such that each K_n is a CM field.

Remark. As Washington observes [Wa 2], if Leopoldt's conjecture is true, then K_∞ is necessarily the cyclotomic $\mathbf{Z}_p$-extension, as follows from Theorem 6.2 of Chapter 5.

The real subfield K_∞^+ is a $\mathbf{Z}_p$-extension of K_0^+. It is then clear that

$$K_\infty = K K_\infty^+,$$

so K_∞ is the lifting over K of the $\mathbf{Z}_p$-extension K_∞^+ of K_0^+.

Let $C_n = C_n^{(p)}$ be the p-primary part of the ideal class group of K_n. We have the orders

$$\operatorname{ord}_p C_n = e_n, \qquad \operatorname{ord}_p C_n^- = e_n^-, \qquad \operatorname{ord}_p C_n^+ = e_n^+,$$

and $e_n = e_n^+ + e_n^-$. Similarly, we have the Iwasawa invariants associated with the Λ-modules

$$C = \varprojlim C_n, \qquad C^- = \varprojlim C_n^-, \qquad C^+ = \varprojlim C_n^+,$$

and we denote these invariants by m, λ with a plus or minus sign as superscript corresponding to the two cases. Then

$$m = m^- + m^+ \quad \text{and} \quad \lambda = \lambda^- + \lambda^+.$$

Indeed, we have for an odd prime p,

$$C = C^- \oplus C^+.$$

Remark. For $p = 2$ one has to be more careful, and for instance, one has to distinguish $C(K^+)$ from its natural image in $C(K)$, which we might denote by C^+. Cf. Lemma 2, and also Lemma 4.1 below.

Theorem 2.1. *Let p be an odd prime. Let K be a CM field, and assume that the p-th roots of unity are in K. Let C be the p-primary part of the ideal class group of K. Then*

(i) $$\operatorname{rank}_p C^+ \leq \operatorname{rank}_p C^- + 1.$$

Let W_K be the group of roots of unity in K. If $K(W_K^{1/p})$ is ramified over K, then

(ii) $$\operatorname{rank}_p C(p)^+ \leq \operatorname{rank}_p C(p)^-.$$

Proof. Let L be the maximal abelian extension of K of exponent p. Let $G = \operatorname{Gal}(L/K)$. By class field theory,

$$G \approx C(p) \quad \text{and} \quad G^+ \approx C(p)^+.$$

Since the p-th roots of unity are assumed to be in K, the extension L is a Kummer extension. Let B be its Kummer group containing K^{*p}, so that

$$K^* \supset B \supset K^{*p} \quad \text{and} \quad L = K(B^{1/p}).$$

We have the exact Kummer duality

$$G \times B/K^{*p} \to \mu_p.$$

Since p is assumed to be odd, we have a direct product decomposition

$$G = G^+ \times G^-.$$

For simplicity of notation, let

$$V = B/K^{*p} = V^+ \times V^-.$$

Since μ_p is a (-1)-eigenspace for complex conjugation, it follows that we have an exact pairing

$$G^+ \times V^- \to \mu_p,$$

and therefore

$$\operatorname{rank}_p G^+ = \operatorname{rank}_p C(p)^+ = \operatorname{rank}_p V^-.$$

On the other hand, if $b \in B$ we know that $K(b^{1/p})$ is unramified, and hence there exists an ideal $\mathfrak{b}$ of K such that $(b) = \mathfrak{b}^p$. The map $b \mapsto \operatorname{Cl}(\mathfrak{b})$ gives rise to a homomorphism

$$\varphi : V = B/K^{*p} \to C_p,$$

which induces a homomorphism

$$\varphi^- : V^- \to C_p^-$$

of V^- into C_p^- because φ commutes with complex conjugation. One then verifies immediately that we obtain an injective map

$$\operatorname{Ker} \varphi^- \to E_K(p)^-,$$

by writing $b \in B$ as $b = a^p u$ with $u \in E_K$ and mapping $b \mapsto u$. Since $(E_K : W_K E_K^+) = 1$ or 2, we obtain

$$E_K(p)^- = W_K(p)^- = W_K(p)$$

because $W_K = W_K^-$. Therefore

$$\operatorname{rank}_p W_K(p) = 1 = \operatorname{rank}_p E_K(p)^-,$$

and

$$\operatorname{rank}_p C(p)^+ = \operatorname{rank}_p V^- \leq 1 + \operatorname{rank}_p C_p^-.$$

This proves the first assertion. Furthermore, if $K(W_K^{1/p})$ is ramified over K, then B does not contain W_K, and consequently

$$\mathrm{Ker}\ \varphi^- = 1.$$

Hence the last inequality on the right can be replaced by the stronger inequality

$$\mathrm{rank}_p\ V^- \le \mathrm{rank}_p\ C^-,$$

thus proving the theorem.

The next two results are really corollaries of the theorem, but we label them theorems in their own right in view of their importance. The first is a classical result of Kummer.

Theorem 2.2. *Let* $K = \mathbf{Q}(\mu_p)$, *and let* h_p *be the class number of* K. *If* h_p^- *is prime to* p, *then* h_p^+ *is prime to* p, *and so* h_p *is prime to* p.

Proof. Obvious, since the p-rank is 0, and the stronger of the two inequalities applies.

Let K_∞ be a $\mathbf{Z}_p$-extension of K_0 such that each K_n is a CM field. Then C, C^+, C^- are modules over the Iwasawa algebra, and thus we have the invariants

$$r_1(C) = r_1(C^+) + r_1(C^-) = r_1^+ + r_1^-.$$

as defined in §1.

Theorem 2.3. *Let* K_∞ *be a* $\mathbf{Z}_p$-extension of K_0 (p odd), *such that each* K_n *is a CM field. Then*

$$r_1^+ \le r_1^-.$$

In particular, if $m^- = 0$ *then* $m^+ = 0$.

Proof. Immediate from Theorem 2.1 and Lemma 3 of §1.

For the prime $p = 2$ the estimates are not as good, but one has a result of Greenberg (cf. Washington [Wa 2]).

Proposition 2.4. *Let* K *be a CM field, and* $C = C(K)$. *Then*

$$\mathrm{rank}_2\ C_{K^+} \le \mathrm{ord}_2|C_K^-| + 1.$$

If K_∞ is a $\mathbf{Z}_2$-extension of K_0 such that each K_n is a CM field, and if $m^- = 0$ then $m^+ = 0$ also.

Proof. Let $r = \operatorname{rank}_2 C_{K^+}$. Then by Lemma 2,

$$2^r = (C_{K^+} : C_{K^+}^2) = (NC_K : NC^+),$$

where C^+ denotes the image of $C(K^+)$ in $C(K)$. This last index divides

$$(C_K : C^+) = \frac{|C_K|}{|C^+|},$$

which by Lemma 2 divides

$$\frac{2|C_K|}{|C(K^+)|} = 2h_K^-.$$

This proves the inequality. The statement about $m^- = 0$ then follows from the structure theorem, cf. Lemma 3 of §1, as for p odd.

§3. The *l*-primary Part in an Extension of Degree Prime to *l*

In this section we prove a lemma of Iwasawa used by Washington [Wa 2]. We begin by a trivial remark.

Lemma 1. *Let F be a number field and K a Galois extension of degree d. Let l be a prime number not dividing d. Let C_F denote the l-primary part of the ideal class group. Then the natural homomorphism $C_F \to C_K$ is injective, the norm $N_{K/F} : C_K \to C_F$ is surjective, and the following conditions are equivalent:*

(i) $C_F = C_K$.
(ii) $\operatorname{rank}_l C_F = \operatorname{rank}_l C_K$.
(iii) *The norm $N_{K/F} : C_K(l) \to C_F(l)$ is an isomorphism.*

Proof. For the first assertion, suppose an ideal $\mathfrak{a}$ of F becomes principal in K, say $\mathfrak{a} = (\alpha)$. Taking the norm yields

$$\mathfrak{a}^d = (N_{K/F}\alpha),$$

and since d is prime to l, it follows that $\mathfrak{a}$ is also principal. Abbreviate the norm by N. Since

$$NC_F = C_F^d = C_F,$$

it follows that the norm is surjective.

It is clear that (i) implies (ii). Assume (ii). Then the norm in (iii) is an isomorphism because it is surjective. It is also a G-isomorphism, where $G = \text{Gal}(K/F)$. Hence G acts trivially on $C_K(l)$, and therefore

$$N = d \cdot \text{identity on } C_K(l).$$

It follows that $C_F(l) = C_K(l)$. Nakayama's lemma or the structure theorem for abelian groups concludes the proof that $C_F = C_K$.

Lemma 2. *Let p be a prime number $\neq l$. Let K_n be a cyclic extension of a number field K_0, of degree p^n. Let f be the order of l mod p^n. Let C_v be the l-primary part of the ideal class group in the subfield of degree p^v. Let D_n be the kernel in the exact sequence*

$$0 \to D_n \to C_n(l) \xrightarrow{\ \text{Norm}\ } C_{n-1}(l) \to 0.$$

Then $D_n \neq 0$ if and only if $C_n \neq C_{n-1}$, and in that case,

$$\dim D_n \geq f.$$

Proof. Let $G = \text{Gal}(K_n/K_0)$. We have a representation of G on the $\mathbf{Z}(l)$-vector space $D = D_n$, and we first show that if $D \neq 0$, then the representation is faithful. If not, it is not injective on the unique cyclic subgroup of order p, namely $\text{Gal}(K_n/K_{n-1})$. Hence it is trivial on that subgroup, and therefore

$$N_{n,n-1}D = pD = 0,$$

whence $D = 0$.

Now assume $D \neq 0$. Then the tensor product of D with the algebraic closure of $\mathbf{F}_l$ splits as a G-direct sum

$$D \otimes \mathbf{F}_l^a = \bigoplus D_i,$$

where the D_i are irreducible components, of dimension 1. If the operation of G on every D_i is not faithful, then it is not faithful on the unique subgroup of order p, so not faithful on D itself. By what we have shown, it follows that G operates faithfully on some D_{i_0}. Hence G operates on this D_{i_0} by a representation into the p^n-th roots of unity, and a generator of G goes to a primitive p^n-th root of unity ζ. But D is defined over $\mathbf{F}_l$, so the conjugate representations by means of the conjugates ζ^l, ζ^{l^2}, ... occur. So exactly f distinct conjugates occur among the D_i, so $\dim D \geq f$. This proves the lemma.

This last part of the argument proves the general statement:

Let D be a finite dimensional representation of a cyclic group G of order p^n, over the prime field $\mathbf{F}_l$. If D is faithful, then

$$\dim D \geq f,$$

where f is the order of l mod p^n.

Washington's application of the lemma then runs as follows.

Theorem 3.1. *Let K_∞ be a $\mathbf{Z}_p$-extension of K_0, and let C_n be the l-primary part of the ideal class group in K_n, where l is a prime $\neq p$. If the l-ranks of C_n are bounded, then the orders of C_n are also bounded for all n.*

Proof. Otherwise, we must have $D_n \neq 0$ for arbitrarily large n, and the order f_n of l mod p^n satisfies

$$f_n \gg p^n.$$

The preceding lemma would then imply that the ranks tend to infinity, a contradiction, which proves the theorem.

Remark. The lemma need not only be applied to a $\mathbf{Z}_p$-extension. For instance, given a number field F, and a prime number l, if K is cyclic of degree p over F, then either the l-rank of C_K tends to infinity, or $C_F = C_K$ as $p \to \infty$. It would be interesting to investigate for what sequence of cyclic extensions of degree p does the l-rank remain bounded, or tends to infinity.

Theorem 3.2. *Let K_∞ be a $\mathbf{Z}_p$-extension of K_0. Assume that each K_n is a CM field. Let l be a prime number $\neq p$, and assume that the l-th roots of unity are in K_0. If $\mathrm{ord}_l|C_n^-|$ is bounded, then $\mathrm{ord}_l|C_n^+|$ is also bounded.*

Proof. By Theorem 2.1 the l-rank of C_n^+ is bounded, so the l-rank of C_n is bounded. Then Theorem 3.1 concludes the proof.

§4. A Relation between Certain Invariants in a Cyclic Extension

The result of this section will be used later to give examples due to Iwasawa, of $\mathbf{Z}_p$-extensions in which orders of ideal class groups tend rapidly to infinity. Results like the first lemma are classical. The prime power case was known in the last century, and the general cyclic case is in Takagi and Chevalley's thesis on class field theory.

Let K be a cyclic extension of a number field F. Let $G = \mathrm{Gal}(K/F)$. For each (normalized) absolute value v of F we let $e(v)$ be the ramification index of v in K. If $v = v_{\mathfrak{p}}$ for some prime ideal $\mathfrak{p}$ of F, then $e(v) = e(\mathfrak{p})$ is the usual ramification index of $\mathfrak{p}$ in K. If v is Archimedean, then $e(v) = 1$ or 2 according as the local extension is trivial or of degree 2 (complex numbers over the real numbers). We put

$$e(K/F) = \prod_v e(v).$$

Then $e(K/F) = e_0(K/F)e_\infty(K/F)$, where

$$e_0(K/F) = \prod_{\mathfrak{p}} e(\mathfrak{p}) \quad \text{and} \quad e_\infty(K/F) = \prod e(v_\infty).$$

We let E denote the group of units and C the group of ideal classes. If G acts on a module A, we let A^G be the submodule of elements fixed under G. The next lemma implies that highly ramified primes have a tendency to generate independent ideal classes, and that the obstruction to this is contained in some cohomology.

Lemma 4.1. *Let K/F be a cyclic extension with Galois group G. Then*

$$|C_K^G| = \frac{h(F)e(K/F)}{[K:F](E_F : N_{K/F}K^* \cap E_F)}.$$

Proof. We assume that the reader is acquainted with a minimum of Galois cohomology for cyclic groups. What is needed is covered for instance in Chapter IX, §1 of my *Algebraic Number Theory*, referred to as *ANT*.

Let I denote the ideal group and P the principal ideal group. The group C_K^G occurs naturally at the beginning of the cohomology sequence associated with the exact sequence

$$0 \to P_K \to I_K \to C_K \to 0.$$

Note that I_K is the direct sum of its semilocal components over primes of F, and by semilocal theory, we have

$$H^1(G, I_K) = \bigoplus_{\mathfrak{p}} H^1(G_{\mathfrak{p}}, \mathbf{Z}),$$

where $G_{\mathfrak{p}}$ is the decomposition group. But $H^1(G_{\mathfrak{p}}, \mathbf{Z}) = 0$, so

$$H^1(G, I_K) = 0.$$

The exact cohomology sequence then yields

$$0 \to P_K^G \to I_K^G \to C_K^G \to H^1(G, P_K) \to 0,$$

whence an exact sequence

$$0 \to I_K^G/P_K^G \to C_K^G \to H^1(G, P_K) \to 0$$

which yields the index relation

(1)
$$|C_K^G| = (I_K^G : P_K^G)|H^1(P_K)|.$$

We analyze the two indices on the right-hand side.
   From the inclusions

$$I_K^G \supset P_K^G \supset P_F$$

we obtain the index

$$(I_K^G : P_K^G) = \frac{(I_K^G : P_F)}{(P_K^G : P_F)}$$

$$= \frac{(I_K^G : I_F)(I_F : P_F)}{(P_K^G : P_F)}$$

(2)
$$= e_0(K/F) \frac{h_F}{(P_K^G : P_F)}.$$

We now use the second exact sequence

$$0 \to E_K \to K^* \to P_K \to 0.$$

We get the beginning of the cohomology sequence

$$0 \to E_F \to F^* \to P_K^G \to H^1(E_K) \to 0$$

because $H^1(K^*) = 0$ by Hilbert's Theorem 90. Hence

$$(P_K^G : P_F) = |H^1(E_K)|$$

$$= |H^0(E_K)| \frac{[K:F]}{e_\infty(K/F)}$$

(by Corollary 2 of Theorem 1, ANT Chapter IX), so by definition,

(3)
$$(P_K^G : P_F) = (E_F : N_{K/F} E_K)[K : F]/e_\infty(K/F).$$

   Once more, from the second exact sequence, we have another portion of the cohomology sequence

$$0 = H^1(K^*) \to H^1(P_K) \to H^0(E_K) \to H^0(K^*).$$

The map from $H^0(E_K)$ to $H^0(K^*)$ is the natural homomorphism

$$E_F/N_{K/F}E_K \to F^*/N_{K/F}K^*.$$

By exactness, we find

$$|H^1(P_K)| = |\mathrm{Ker}(E_F/N_{K/F}E_K \to F^*/N_{K/F}K^*)|$$
$$(4) \qquad\qquad = (N_{K/F}K^* \cap E_F : N_{K/F}E_K).$$

Using the inclusions

$$E_F \supset (N_{K/F}K^* \cap E_F) \supset N_{K/F}E_K,$$

and putting (1), (2), (3), (4) together proves the lemma.

We shall apply the lemma as does Iwasawa [Iw 15] to get an example of a field with a highly divisible class number.

Lemma 4.2. *Let l be an integer ≥ 2. Let K_d be an extension of a number field K of degree d. Let $q_1, \ldots, q_t$ be prime ideals of K which split completely in K_d. Let K' be a cyclic extension of K, of degree l, in which $q_1, \ldots, q_t$ are totally ramified. Let $K'_d = K'K_d$. Then*

$$\frac{|C(K'_d)|}{|C(K_d)|} \quad \text{is divisible by} \quad l^{(t-[K:\mathbf{Q}])d-1}.$$

Proof. We have the diagram

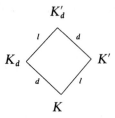

The extensions K_d and K' are linearly disjoint because of the way any one of the primes q_i splits in them. Thus K'_d has degree l over K_d. We apply Lemma 4.1 to the cyclic extension K'_d over K_d. This yields:

$$\frac{|C(K'_d)|}{|C(K_d)|} \quad \text{is divisible by} \quad \frac{e(K'_d/K_d)}{[K'_d:K_d](E_d:E'_d)},$$

where $E_d = E(K_d)$ is the group of units of K_d. Indeed, E_d^l is contained in the norm group from K_d'. All we have to do is now estimate the divisibility of the expression on the right.

Each ideal q_i splits into d primes in K_d, and each factor in K_d is then totally ramified in K_d'. Consequently

$$e(K_d'/K_d) \quad \text{is divisible by} \quad l^{td}.$$

On the other hand,

$$[K_d' : K_d] = l.$$

Thirdly, E_d mod roots of unity has rank bounded by

$$[K_d : \mathbf{Q}] - 1 = d[K : \mathbf{Q}] - 1,$$

and consequently

$$(E_d : E_d^l) \text{ divides } l^{d[K:\mathbf{Q}]}.$$

Putting these three estimates together proves the lemma.

§5. Examples of Iwasawa

Let K be a number field and let K_∞ be a $\mathbf{Z}_p$-extension. Let q be a prime ideal of K. The decomposition group of q in $\Gamma = \text{Gal}(K_\infty/K)$ is either trivial or closed of finite index. In the first case, we say that q **splits completely**, and in the second case, we say that q is **finitely decomposed** in K_∞.

Let l be a prime number, which may be equal to p. We are interested in giving examples of Iwasawa [Iw 15] for which the l-primary part of the ideal class group C_n grows exponentially. We shall use the formula of the last section, applied to extensions where the ramification indices grow faster than the unit index in the denominator, times the degree. This implies that the class number on the left-hand side of the relation grows equally rapidly.

Let K' be a finite extension of K. Then $K_\infty K' = K_\infty'$ is a $\mathbf{Z}_p$-extension of K'. If q splits completely in K_∞, then any divisor q' of q in K' splits completely in K_∞'. This is an elementary fact of algebraic number theory. In particular, if $q_1', \ldots, q_t'$ in K' lie above t such primes $q_1, \ldots, q_t$, they will be distinct primes of K' splitting completely in K_∞'.

Theorem 5.1. *Let K_∞/K be a $\mathbf{Z}_p$-extension. Let $q_1, \ldots, q_t$ be prime ideals of K which split completely in K_∞. Let K' be a cyclic extension of K of degree l, in which $q_1, \ldots, q_t$ are totally ramified. Then*

$$\text{ord}_l |C(K_n')| \geq (t - [K:\mathbf{Q}])p^n - 1.$$

Proof. This is merely a special case of Lemma 4.2.

A concrete example of the situation as in the theorem can be given once we have shown below how to construct a $\mathbf{Z}_p$-extension in which infinitely many primes of K split completely. Then we can take t arbitrarily large. We can then first lift the extension over the field $K(\boldsymbol{\mu}_l)$, and thus assume without loss of generality that K contains the l-th roots of unity. We then select an element α of K divisible exactly by the first power of $q_1, \ldots, q_t$ (and possibly other primes). We let

$$K' = K(\alpha^{1/l}).$$

Then the hypotheses of the theorem are satisfied.

Theorem 5.2. *Let K be a CM field. Then:*

(i) *There exists a $\mathbf{Z}_p$-extension K_∞ of K, Galois over K^+, such that if $\Gamma = \mathrm{Gal}(K_\infty/K)$, then*

$$\Gamma = \Gamma^-.$$

(ii) *For any such extension, let q^+ be a prime ideal of K^+ which does not divide p and remains prime in K, so q is its unique extension in K. Then q splits completely in K_∞. (Tchebotarev guarantees that there exist infinitely many such primes.)*

Proof. Let $M_p(K)$ be the maximal p-abelian p-ramified extension of K. By class field theory, e.g. Chapter 5, §5, there is a quasi-isomorphism

$$\mathrm{Gal}(M_p(K)/K) \sim U_p/\bar{E},$$

where U_p is the product of the local unit groups $U_{\mathfrak{p}}$ at the primes $\mathfrak{p}$ above p, and $\bar{E}$ is the closure of the global units in U_p. Each real prime $\mathfrak{p}^+$ either remains prime in K or splits into two primes $\mathfrak{p}_1, \mathfrak{p}_2$ in K. In either case, the semilocal component

$$\prod_{\mathfrak{p}|\mathfrak{p}^+} U_{\mathfrak{p}}$$

of U_p contains a subgroup of finite index isomorphic under the exponential map with

$$\prod_{\mathfrak{p}|\mathfrak{p}^+} p^m \mathfrak{o}_{\mathfrak{p}},$$

for m sufficiently large. Here $\mathfrak{o}_{\mathfrak{p}}$ denotes the local ring of integers at $\mathfrak{p}$. From this it is clear that U_p^- has $\mathbf{Z}_p$-rank ≥ 1. Furthermore, E contains a subgroup of finite index which is real, so $\bar{E}$ contains a subgroup of finite index which is fixed under complex conjugation. Put

$$\mathscr{G} = \mathrm{Gal}(M_p(K)/K),$$

and for simplicity of language, assume that p is odd. Then

$$\mathscr{G} = \mathscr{G}^+ \times \mathscr{G}^-,$$

and the $\mathbf{Z}_p$-rank of $\mathscr{G}^-$ is ≥ 1. Hence there exists a factor group Γ of $\mathscr{G}$ such that $\Gamma = \Gamma^-$, and Γ is isomorphic to $\mathbf{Z}_p$ as compact group. By Galois theory, Γ is the Galois group of a $\mathbf{Z}_p$-extension K_∞ of K, which is normal over K^+.

Let $\mathfrak{q}^+$ be a prime ideal of K^+ which remains prime in K. Let D be the decomposition group of $\mathfrak{q}^+$ in the group

$$\mathscr{G} = \mathrm{Gal}(K_\infty/K^+).$$

Then D is (topologically) cyclic because $\mathfrak{q}$ is unramified in K_∞. But $\mathscr{G}$ is "dihedral," in other words, $\mathscr{G}$ is generated by complex conjugation and Γ satisfying the relations

$$\rho\gamma\rho = \gamma^{-1},$$

for $\gamma \in \Gamma$. Since $\mathfrak{q}^+$ remains prime in K, we cannot have $D \subset \Gamma$. Then D contains an element $\rho\gamma$, and any such element has order 2. It is then immediately verified that D is cyclic of order 2 and that its intersection with Γ is 1. This proves that $\mathfrak{q}$ splits completely in K_∞, and concludes the proof.

Remark. If $p = 2$, then one has to take a factor group of $\mathscr{G}$ by $\mathscr{G}^{1+\rho}$ to obtain the minus part, and the argument is essentially the same.

§6. A Lemma of Kummer

Theorem 6.1. *Let p be an odd prime. Let u be a unit in $\mathbf{Q}(\boldsymbol{\mu}_p)$. Suppose there exists an integer $a \in \mathbf{Z}$ such that $u \equiv a \pmod{p}$. If p does not divide the class number h_p, then u is a p-th power in $\mathbf{Q}(\boldsymbol{\mu}_p)$.*

Proof. Let $K = \mathbf{Q}(\boldsymbol{\mu}_p)$. By class field theory, it suffices to show that the extension $K(u^{1/p})$ is unramified, because the hypothesis then implies that $K(u^{1/p}) = K$, so u is a p-th power in K. Raising u to the $(p - 1)$-th power allows us to assume without loss of generality that $u \equiv 1 \pmod{p}$.

We contend that

$$u \equiv 1 \pmod{\pi p},$$

where $\pi = \zeta - 1$ and ζ is a primitive p-th root of unity. Otherwise, we have

$$u \equiv 1 + px \bmod \pi p$$

with x equal to a p-unit in $\mathbf{Z}_p^*$. But u is a global unit, so

$$N_{K/\mathbf{Q}}(u) = \pm 1 \equiv (1 + px)^{p-1} \equiv 1 - px \pmod{\pi p}.$$

In both cases where the norm is 1 or -1 we get a contradiction.

Let $\alpha = u^{1/p}$ be any p-th root of u. Then

$$\frac{1 - \alpha}{\pi} \quad \text{is a root of} \quad \frac{(\pi X - 1)^p + u}{\pi^p},$$

and this polynomial has p-integral coefficients. Its other roots are

$$\frac{1 - \zeta^i \alpha}{\pi},$$

and the different is a product of terms

$$\frac{(\zeta^i - \zeta^j)\alpha}{\pi},$$

each of which is a p-unit. Hence $K(\alpha) = K(u^{1/p})$ is unramified at p, and is trivially unramified at all other primes. This concludes the proof.

p-adic Preliminaries 14

The first section introduces the p-adic gamma function of Morita. After that, we deal with a topic which can be viewed temporarily as independent, the Artin–Hasse power series, and the Dwork power series closely related to it. The latter allows us to obtain an analytic representation of p-th roots of unity, which reappear later in the context of gauss sums, occurring as eigenvalues of p-adic completely continuous operators. Cf. Dwork's papers in the bibliography.

§1. The p-adic Gamma Function

In this section, we give Morita's definition of the p-adic gamma function. To start, we let f be the function defined for positive integers n by the formula

$$f(n) = (-1)^n \prod_{\substack{j=1 \\ (p,\, j)=1}}^{n} j.$$

We wish to show that there exists a continuous function on $\mathbf{Z}_p$ which restricts to f on the positive integers $\mathbf{Z}^+$. For this purpose, it suffices to prove the following lemma.

Lemma 1.1. *For any positive integers N, n, k we have*

$$f(n + p^N k) \equiv f(n) \pmod{p^N}.$$

Proof. Let $G = \mathbf{Z}(p^N)^*$. Pairing an element and its inverse in G we find:

$$\prod_{j \in G} j \equiv \begin{cases} -1 & \bmod\ p^N \text{ if } p \text{ is odd} \\ 1 & \bmod\ p^N \text{ if } p = 2. \end{cases}$$

If p is odd, then

$$\frac{f(n + p^N k)}{f(n)} = (-1)^{p^N k} \prod_{\substack{n+1 \le j \le n + p^N k \\ (j,\, p) = 1}} j$$

$$\equiv (-1)^k \left(\prod_{j \in G} j \right)^k \equiv 1 \bmod p^N,$$

as desired. The proof is similar for $p = 2$ and is left to the reader.

By the lemma, we can extend f by continuity to all of $\mathbf{Z}_p$, and since $f(n)$ is a p-adic unit for n equal to a positive integer, it follows that

$$f : \mathbf{Z}_p \to \mathbf{Z}_p^*$$

is a map of $\mathbf{Z}_p$ into $\mathbf{Z}_p^*$.

We define the **p-adic gamma function**

$$\Gamma_p(x) = -f(x - 1).$$

Thus on integers $n \ge 2$ we have

$$\boxed{\Gamma_p(n) = (-1)^n \prod_{\substack{j = 1 \\ (p,\, j) = 1}}^{n-1} j.}$$

We shall write Γ instead of Γ_p when p is fixed, and in particular, for the rest of this section.

Let us now calculate a few values, especially of negative integers. If $u(x)$, $v(x)$ are continuous functions of a p-adic variable x, we define:

$$\begin{cases} u(x) \\ v(x) \end{cases} \text{ is the function such that } \quad x \mapsto \begin{cases} u(x) & \text{if } x \in \mathbf{Z}_p^* \\ v(x) & \text{if } x \in p\mathbf{Z}_p. \end{cases}$$

With this notation, we have

$$\Gamma(n + 1) = \begin{cases} -n\Gamma(n) \\ -\Gamma(n), \end{cases}$$

if n is an integer ≥ 2, and consequently by continuity,

$$\Gamma(x + 1) = \begin{cases} -x\Gamma(x) \\ -\Gamma(x) \end{cases} \quad \text{for all } x \in \mathbf{Z}_p.$$

We call this the **functional equation**.

From $\Gamma(2) = \Gamma(1 + 1) = -\Gamma(1)$, we get

$$\Gamma(1) = -1.$$

From $\Gamma(1) = \Gamma(1 + 0) = -\Gamma(0)$, we get

$$\Gamma(0) = 1.$$

Theorem 1.2. *For any integer $n \geq 1$, we have*

$$\Gamma(n)\Gamma(1 - n) = (-1)^{n + [(n-1)/p]},$$

where the bracket, as usual, is the greatest integer function.

Proof. The theorem is true for $n = 1$ by the above. We can then proceed inductively, using $\Gamma(1 - n) = \{^n_{-1} \Gamma(-n)$, to get:

$$\Gamma(0) = \left(\prod_{\substack{j=1 \\ (j, p) = 1}}^{n-1} j \right)(-1)^{\delta(n-1)}\Gamma(1 - n),$$

where $\delta(n - 1) =$ number of elements among $1, \ldots, n - 1$ which are divisible by p. This is immediate from the functional equation. Since $\delta(n - 1) = [(n - 1)/p]$, the right-hand side is equal to

$$(-1)^n\Gamma(n)(-1)^{[(n-1)/p]}\Gamma(1 - n).$$

The formula of the theorem then follows at once.

Let $x \in \mathbf{Z}_p$. We denote *for p odd*:

$$R(x) = \text{representative of } x \bmod p \text{ in the set } \{1, \ldots, p\}.$$

If p must be in the notation, we would write $R_p(x)$.

Theorem 1.3. *If $p \neq 2$ then*

$$\Gamma(x)\Gamma(1 - x) = (-1)^{R(x)}.$$

If $p = 2$, then

$$\Gamma(x)\Gamma(1 - x) = \varepsilon(x),$$

where

$$\varepsilon(x) = \begin{cases} -1 & \text{if } x \equiv 0, 1 \bmod 4 \\ 1 & \text{if } x \equiv 2, 3 \bmod 4. \end{cases}$$

Proof. By continuity, it suffices to prove the theorem when x is an integer $n \geq 1$. Write

$$n = a_0 + a_1 p + a_2 p^2 + \cdots + a_r p^r,$$

with $a_0 \in \{1, \ldots, p\}$ and $a_i \in \{0, \ldots, p - 1\}$ for $i \geq 1$. Then

$$\left[\frac{n-1}{p} \right] = a_1 + \cdots + a_r p^{r-1}.$$

$$n + \left[\frac{n-1}{p} \right] = a_0 + a_1(1 + p) + a_2(p + p^2) + \cdots + a_r(p^{r-1} + p^r).$$

Hence for p odd,

$$n + \left[\frac{n-1}{p} \right] \equiv a_0 = R(n) \bmod 2.$$

This proves the theorem in the present case.

If $p = 2$, then the parity of $n + [(n - 1)/2]$ does not change for the elements in the residue class of $n \bmod 4$. We then determine explicitly the values of this number for $n = 1, 2, 3, 4$ to get the desired answer.

If $p = 2$, then it is convenient to define

$$R_2(x) = R(x) = \text{representative of } x \bmod 2 \text{ in } \{0, 1\}.$$

In both cases, p odd or even, we define

$$R'(x) = \frac{x - R(x)}{p}.$$

If $p \neq 2$ note that for a positive integer n,

$$R'(n) = \left[\frac{n-1}{p} \right].$$

This follows from the p-adic expansion in the proof of Theorem 1.3.

Theorem 1.4 (Distribution relation). *Let N be an integer ≥ 2 and prime to p. Then*

$$\prod_{i=0}^{N-1} \Gamma\left(\frac{x + i}{N} \right) = \Gamma(x) \prod_{i=1}^{N-1} \Gamma\left(\frac{i}{N} \right) g_N(x)^{-1},$$

where $g_N(x)$ is given by the formulas:

$$g_N(x) = N^{R(x)-1}N^{(p-1)R'(x)} \quad \text{if } p \neq 2$$
$$g_N(x) = N^{R'(x)} \quad\qquad \text{if } p = 2.$$

Remark. Both $R(x)$ and $R'(x)$ are continuous functions of $x \in \mathbf{Z}_p$. Since $R(x)$ is a positive integer, the exponentiation $N^{R(x)}$ is well defined. Since $N^{p-1} \equiv 1 \pmod{p}$, its exponentiation with $R'(x)$ is also well defined. When $x = n$ is a positive integer, then we can simplify the formula for $g_N(n)$, using

$$R(n) + pR'(n) = n.$$

For instance, for p odd, $g_N(n) = N^{n-1-R'(n)}$.

Proof. Define $g_N(x)$ by using the relation to be proved, so

$$g_N(x) = \prod_{i=1}^{N-1} \Gamma\left(\frac{i}{N}\right) \prod_{i=0}^{N-1} \Gamma\left(\frac{x+i}{N}\right)^{-1} \Gamma(x).$$

By continuity, it suffices to prove the theorem when x is a positive integer. We have

$$g_N(0) = 1$$

$$\frac{g_N(x+1)}{g_N(x)} = \frac{\Gamma(x+1)}{\Gamma(x)} \frac{\Gamma\left(\dfrac{x}{N}\right)}{\Gamma\left(\dfrac{x}{N}+1\right)}.$$

From the functional equation, we get

$$g_N(x+1) = \frac{\begin{cases} -x \\ -1 \end{cases}}{\begin{cases} -x/N \\ -1 \end{cases}} g_N(x) = \begin{cases} N \\ 1 \end{cases} g_N(x).$$

Hence for every positive integer n we find

$$g_N(n) = N^{n-1-\delta(n-1)} = N^{n-1-[(n-1)/p]}.$$

If p is odd, then $[(n-1)/p] = R'(n)$ and the assertion follows. The same argument also works for $p = 2$ and is left to the reader.

The next result shows that the fudge product occurring in the distribution relation is a fourth root of unity.

Proposition 1.5.

$$\prod_{i=1}^{N-1} \Gamma\left(\frac{i}{N}\right) = \begin{cases} \pm 1, & \text{if } N \text{ is odd} \\ \pm 1, \pm\sqrt{-1} & \text{if } N \text{ is even.} \end{cases}$$

Proof. Suppose N is odd. We write

$$\prod_{i=1}^{N-1} \Gamma\left(\frac{i}{N}\right) = \prod_{i=1}^{(N-1)/2} \Gamma\left(\frac{i}{N}\right)\Gamma\left(1 - \frac{i}{N}\right),$$

and each factor on the right is ± 1 by Theorem 1.3. If N is even, then

$$\prod_{i=1}^{N-1} \Gamma\left(\frac{i}{N}\right) = \Gamma\left(\frac{1}{2}\right) \prod_{i \le (N-1)/2} \Gamma\left(\frac{i}{N}\right)\Gamma\left(1 - \frac{i}{N}\right).$$

Furthermore,

$$\Gamma(\tfrac{1}{2})\Gamma(\tfrac{1}{2}) = \pm 1,$$

and the proposition follows.

§2. The Artin–Hasse Power Series

Let p be a prime number. Define the **Artin–Hasse power series** by

$$\text{AH}(X) = \exp\left(\sum_{n=0}^{\infty} \frac{X^{p^n}}{p^n}\right).$$

As usual, exp is the standard power series for the exponential function. Then $\text{AH}(X)$ has rational coefficients.

Theorem 2.1. *We have* $\text{AH}(X) \in \mathbf{Z}_p[[X]]$.

Lemma (Dieudonné–Dwork). *Let* $f(X) \in 1 + X\mathbf{Q}_p[[X]]$. *Then*

$$f(X) \in 1 + X\mathbf{Z}_p[[X]] \quad \Leftrightarrow \quad \frac{f(X)^p}{f(X^p)} \in 1 + pX\mathbf{Z}_p[[X]].$$

Proof. The left-hand side obviously implies the right-hand side in the equivalence to be proved. So assume the right-hand side. If R is any ring, and

$$f(X) \in 1 + XR[[X]],$$

then a simple recursion shows that $f(X)$ has an infinite product expression

$$f(X) = \prod_{n=1}^{\infty} (1 - a_n X^n) \quad \text{with } a_n \in R.$$

Furthermore, the elements a_n are uniquely determined.

Assume that $f(X)^p / f(X^p) \in 1 + pX\mathbf{Z}_p[[X]]$. Suppose that some coefficient a_n is not p-integral. Without loss of generality, we may assume that

$$f(X) = \prod_{n=r}^{\infty} (1 - a_n X^n) = 1 - a_r X^r + \text{higher terms}$$

and $a_r \notin \mathbf{Z}_p$. Then

$$\frac{f(X)^p}{f(X^p)} = \frac{1 - pa_r X^r + \cdots}{1 - a_r X^{rp} + \cdots} = 1 - pa_r X^r + \text{higher terms}.$$

By assumption, we must have $pa_r \in p\mathbf{Z}_p$ so $a_r \in \mathbf{Z}_p$, which proves the lemma.

We apply the test of the lemma to the Artin–Hasse series. We thus find

$$\frac{AH(X)^p}{AH(X^p)} = \frac{\exp(p \sum X^{p^n}/p^n)}{\exp(\sum X^{p^{n+1}}/p^n)} = \exp(pX) = \sum \frac{p^n}{n!} X^n.$$

To apply the lemma it suffices that ord $p^n/n! \geq 1$ for all n, which is the case. This concludes the proof of the theorem.

Let us write

$$\exp\left(x + \frac{x^p}{p}\right) = \sum_{k=0}^{\infty} c_k x^k.$$

We wish to give estimates for the coefficients c_k. Let

$$\text{ord} = \text{ord}_p.$$

We shall prove:

(1)
$$\text{ord } c_k \geq -\frac{k}{p^2}\left(2 + \frac{1}{p-1}\right).$$

Proof. We can write

$$\exp\left(x + \frac{x^p}{p}\right) = AH(x) \prod_{n \geq 2} \exp\left(-\frac{x^{p^n}}{p^n}\right).$$

Write

$$\exp\left(-\frac{x^{p^n}}{p^n}\right) = \sum_{k=0}^{\infty} d_k^{(n)} x^k = \sum_{m=0}^{\infty} \frac{x^{p^n m}}{p^{nm} m!} (-1)^m.$$

We prove:

(2) $$\operatorname{ord} d_k^{(n)} \geq -\frac{k}{p^n}\left(n + \frac{1}{p-1}\right).$$

Indeed, for any positive integer m, we recall that

$$\operatorname{ord} m! = \frac{m - s(m)}{p - 1},$$

where $s(m) = s_p(m)$ is the sum of the coefficients in the standard p-adic expansion of m. Hence

$$\operatorname{ord} d_{mp^n}^{(n)} = -nm - \frac{m - s(m)}{p - 1} \geq -nm - \frac{m}{p - 1}.$$

Factoring out $k = mp^n$ immediately implies (2).
    Since

$$\min_{n \geq 2} \frac{-1}{p^n}\left(n + \frac{1}{p-1}\right) = \frac{-1}{p^2}\left(2 + \frac{1}{p-1}\right),$$

it follows that

(3) $$\operatorname{ord} d_k^{(n)} \geq \frac{-k}{p^2}\left(2 + \frac{1}{p-1}\right).$$

Hence the coefficients in the power series expansion of the product

$$AH(x) \prod_{n=2}^{\infty} \exp\left(-\frac{x^{p^n}}{p^n}\right) = \exp\left(x + \frac{x^p}{p}\right)$$

satisfy an estimate like (3), and therefore the coefficients c_k satisfy

$$\text{ord } c_k \geq -\frac{k}{p^2}\left(2 + \frac{1}{p-1}\right).$$

This proves the desired estimate (1).

For each element π such that $\pi^{p-1} = -p$, we define **Dwork's power series**

$$E_\pi(X) = \exp(\pi X - \pi X^p).$$

This can also be written

$$E_\pi(X) = \exp\left(\pi X + \frac{(\pi X)^p}{p}\right) = \sum e_n X^n.$$

The coefficients e_n lie in $\mathbf{Q}_p(\pi)$. In particular

$$\text{ord } e_n \in \frac{1}{p-1}\,\mathbf{Z}.$$

As usual, ord $=$ ord$_p$. If we want ord$_\pi$, then we shall specify π in the notation. Of course,

$$\text{ord}_\pi = (p-1)\text{ord}_p.$$

Lemma 2.2.

(i) *We have* ord $e_n \geq n(p-1)/p^2$, *and* e_n *is p-integral.*

(ii) *If* $n \geq 2$ *then* ord$_\pi e_n \geq 2$.

Proof. By (1), letting $X \mapsto \pi X$, we find at once that

$$\text{ord } e_n \geq n\left(\frac{p-1}{p^2}\right).$$

To prove (ii), i.e. to prove that

$$\text{ord } e_n \geq \frac{2}{p-1} \quad \text{for } n \geq 2,$$

it suffices to show that

$$\text{ord } e_n > \frac{1}{p-1}$$

because ord e_n is a fraction whose denominator is $p - 1$. The inequality for ord e_n follows if

$$n\left(\frac{p-1}{p^2}\right) > \frac{1}{p-1}, \quad \text{or equivalently,} \quad n > \left(\frac{p}{p-1}\right)^2.$$

For $p \geq 5$ or $n \geq 5$ there is no problem. For $p = 3$ and $n \geq 3$, there is also no problem. For $p = 3$ and $n = 2$, we get directly

$$\exp(\pi X - \pi X^p) = 1 + \pi X + \tfrac{1}{2}\pi^2 X^2 + \text{higher terms},$$

so we get the right lower bound for ord e_2. For $p = 2$, we just compute explicitly the coefficients of X^2, X^3, X^4 and find that they are divisible by 4, thus settling the final cases, and proving the lemma.

Remark. From the first part of the lemma, we know that the coefficients of $E_\pi(X)$ tend to 0. In particular, we can evaluate $E_\pi(X)$ by substituting any p-adic number for X in the power series if this p-adic number has absolute value ≤ 1. However this value cannot usually be found by substituting the number directly in the expression $\exp(\pi X - \pi X^p)$. We shall see an example of this in the next section.

§3. Analytic Representation of Roots of Unity

Let p be a prime number. Throughout this section, we let $\pi \in \mathbf{C}_p$ be an element such that

$$\pi^{p-1} = -p.$$

Lemma 3.1. *For each element $\pi \in \mathbf{C}_p$ such that $\pi^{p-1} = -p$, there exists a unique p-th root of unity ζ_π such that*

$$\zeta_\pi \equiv 1 + \pi \pmod{\pi^2}.$$

The correspondence $\pi \mapsto \zeta_\pi$ establishes a bijection between p-th roots of unity $\neq 1$, and elements π as above.

Proof. If ζ is a non-trivial p-th root of unity, then $(\zeta - 1)^{p-1} \sim p$ (where $x \sim y$ means that x/y is a p-unit). If ζ_1, ζ_2 are two primitive p-th roots of unity which are both $\equiv 1 + \pi \pmod{\pi^2}$, then $\zeta_1 - \zeta_2 \equiv 0 \pmod{\pi^2}$, whence $\zeta_1 = \zeta_2$. Since there are exactly $p - 1$ non-trivial p-th roots of unity, and $p - 1$ possible elements π in $\mathbf{Q}_p(\pi) = \mathbf{Q}_p(\zeta)$, and since any element $\xi \equiv 1 \pmod{\pi}$ but $\xi \not\equiv 1 \pmod{\pi^2}$ can be written in the form

$$\xi \equiv 1 + \gamma\pi \pmod{\pi^2}$$

for some $(p - 1)$th root of unity γ, both existence and the bijection follow.

Remark. Each root of unity ζ gives rise to a character

$$\psi : \mathbf{Z}_p \to \boldsymbol{\mu}_p$$

such that $\psi(1) = \zeta$. The unique character whose value at 1 is ζ_π will be denoted by ψ_π, so by definition we have the formula

$$\psi_\pi(1) = \zeta_\pi,$$

and $\psi_\pi(1)$ is the unique root of unity $\equiv 1 + \pi \bmod \pi^2$.

Theorem 3.2 (Dwork). *We have* $E_\pi(1) = \zeta_\pi$, *so* $E_\pi(1)$ *is the unique p-th root of unity* $\equiv 1 + \pi \bmod \pi^2$. *For any* $c \in \mathbf{Z}_p$ *such that* $c^p = c$, *we have*

$$E_\pi(c) = E_\pi(1)^c.$$

Proof. First we observe that $E_\pi(1)$ is defined by substituting 1 for X in the power series for $E_\pi(X)$, which converges in light of the lower bound for the orders of the coefficients. Then note that

$$E_\pi(X)^p = \exp(p\pi X - p\pi X^p) = \exp(p\pi X)\exp(-p\pi X^p)$$

because the factor p inside the exponent makes the two series on the right-hand side converge. Substituting 1 for X can now be done to see at once that $E_\pi(1)^p = 1$. To conclude the proof of the first assertion, it suffices to show that

$$E_\pi(1) \equiv 1 + \pi \,(\bmod \pi^2).$$

This is clear from Lemma 2.2(ii). Similarly,

$$E_\pi(c)^p = \exp(p\pi(c - c^p)) = \exp(0) = 1,$$

and

$$E_\pi(c) \equiv 1 + c\pi \,(\bmod \pi^2),$$

so $E_\pi(c) = E_\pi(1)^c$, thus proving the theorem.

Similarly, let $\zeta \in \boldsymbol{\mu}_{q-1}$ where $q = p^r$ is a power of p. Let $\mathbf{T}$ denote the absolute trace to $\mathbf{F}_p$, and let

$$\psi_{\pi, q} = \psi_\pi \circ \mathbf{T}.$$

Then

$$\psi_{\pi, q} : \mathbf{F}_q \to \boldsymbol{\mu}_p$$

is a character on the additive group $\mathbf{F}_q$, and we let

$$E_{\pi,q}(x) = \exp(\pi x - \pi x^q) = E_\pi(x)E_\pi(x^p)\cdots E_\pi(x^{p^{r-1}}).$$

Theorem 3.3 (Dwork). *We have*

$$E_{\pi,q}(\zeta) = \psi_{\pi,q}(\zeta \bmod p),$$

and this is the unique p-th *root of unity* $\equiv 1 + T(\zeta)\pi \bmod \pi^2$.

Proof. From Lemma 2.2(ii), we know that $E_\pi(\zeta) \equiv 1 + \zeta\pi \bmod \pi^2$. Hence

$$\begin{aligned} E_{\pi,q}(\zeta) &= E_\pi(\zeta)E_\pi(\zeta^2)\cdots E_\pi(\zeta^{p^{r-1}}) \\ &\equiv 1 + T(\zeta)\pi \bmod \pi^2. \end{aligned}$$

The same argument as in Theorem 3.2 shows that $E_{\pi,q}(\zeta)$ is the unique *p*-th root of unity satisfying the above congruence, thus proving the theorem.

Appendix: Barsky's Existence Proof for the *p*-adic Gamma Function

We include this appendix to illustrate techniques which might be useful in similar contexts, say for *p*-adic differential equations, where an ad hoc argument as in §1 cannot be given. No use will be made of this appendix elsewhere in the book.

We wish to show the existence of a continuous function of $\mathbf{Z}_p$ into itself which takes on values related to the factorials at the positive integers. This is an interpolation problem, and thus we begin with a criterion for the existence of a continuous function taking given values. As in Chapter 4, we consider the space of power series

$$g(x) = \sum b_n \frac{x^n}{n!}, \qquad b_n \in \mathbf{C}_p,$$

with $\lim b_n = 0$ (the limit is of course the *p*-adic limit). This space is called the Leopoldt space. It is in fact a Banach algebra, under the sup norm of the coefficients, $\|g\|_{\mathscr{L}} = \max|b_n|$. If *g*, *h* are in the space, so is the product *gh* and

$$\|gh\|_{\mathscr{L}} \leq \|g\|_{\mathscr{L}}\|h\|_{\mathscr{L}}.$$

These properties are trivially verified. Note that a power series

$$\sum b'_n x^n$$

with coefficients b_n' which are p-integral is in the Leopoldt space, as one sees by writing $b_n = b_n' n!$.

Theorem. *Let $\{a_n\}$ be a sequence in $\mathbf{C}_p$. There exists a continuous function*

$$f : \mathbf{Z}_p \to \mathbf{C}_p$$

such that $f(n) = a_n$ for all integers $n \geq 0$ if and only if the following condition is satisfied. Let

$$e^{-x} \sum_{n=0}^{\infty} a_n \frac{x^n}{n!} = \sum_{n=0}^{\infty} b_n \frac{x^n}{n!}.$$

Then $\lim b_n = 0$ (p-adically, of course).

Proof. We use Mahler's theorem (Chapter 4, Theorem 1.3). Any continuous function f has an expansion

$$f(x) = \sum_{n=0}^{\infty} b_n \binom{x}{n} \quad \text{with} \quad b_n \to 0,$$

and

$$b_n = (\Delta^n f)(0).$$

Conversely, if $f : \mathbf{Z}^+ \to \mathbf{C}_p$ is a function such that $\Delta^n f(0) \to 0$ as $n \to \infty$, then f can be extended to a continuous function on $\mathbf{Z}_p$. (We denote by $\mathbf{Z}^+$ the set of integers ≥ 0.) Note that

$$\Delta^n f(0) = \sum_{i=0}^{n} (-1)^{n-i} \binom{n}{i} f(i).$$

The theorem is then obvious in view of the identity

$$e^{-x} \sum f(n) \frac{x^n}{n!} = \sum \Delta^n f(0) \frac{x^n}{n!}.$$

We apply the theorem to the function f defined by

$$f(0) = 1, \qquad f(n) = (-1)^n \prod_{\substack{1 \leq j \leq n \\ (j,\, p) = 1}} j.$$

83

Except for the power of -1, $f(n)$ is equal to $n!$ from which all the factors divisible by p have been deleted. Thus

$$f : \mathbf{Z}^+ \rightarrow \mathbf{Z}_p^*$$

takes its values in *p*-adic units.

This function has a continuous extension to $\mathbf{Z}_p$.

Proof. We have

$$\sum f(n) \frac{x^n}{n!} = \sum_{i=0}^{p-1} \sum_{n=0}^{\infty} f(pn + i) \frac{x^{pn+i}}{(pn + i)!}$$

$$= \sum_{i=0}^{p-1} \sum_{n=0}^{\infty} (-1)^{pn+i} \frac{(pn + i)!}{p^n n!} \frac{x^{pn+i}}{(pn + i)!}$$

$$= \exp\left(\frac{(-x)^p}{p}\right) \sum_{i=0}^{p-1} (-x)^i.$$

Multiplying by e^{-x}, we obtain

$$e^{-x} \sum f(n) \frac{x^n}{n!} = \exp\left(-x + \frac{(-x)^p}{p}\right) \sum_{i=0}^{p-1} (-x)^i$$

$$= \sum b_n \frac{x^n}{n!},$$

with some coefficients b_n, which we must show tend to 0. The set of power series

$$\sum c_n' \frac{x^n}{n!} \quad \text{with } c_n' \rightarrow 0$$

is closed under multiplication. Hence after replacing x by $-x$, it suffices to prove that the coefficients c_k' of the series

$$\exp\left(x + \frac{x^p}{p}\right) = \sum c_k' \frac{x^k}{k!}$$

tend to 0. But by (1) of §2, we know that

$$\operatorname{ord} c_k' \geq \operatorname{ord} k! - \frac{k}{p^2}\left(2 + \frac{1}{p-1}\right)$$

$$= \frac{k - s(k)}{p - 1} - \frac{k}{p^2}\left(2 + \frac{1}{p-1}\right).$$

Since

$$s(k) \le (p - 1)(1 + \log_p k),$$

(where $\log_p$ = high-school-log-to-the-base-p), we obtain

$$\operatorname{ord} c_k' \ge k\left(\frac{1}{p - 1} - \frac{1}{p^2}\left(2 + \frac{1}{p - 1}\right)\right) - 1 - \log_p k,$$

from which it is clear that $c_k' \to 0$. This concludes the proof.

15 The Gamma Function and Gauss Sums

The history of Gauss sums as eigenvalues of Frobenius on Fermat or Artin–Schreier curves goes back to Davenport–Hasse [Da–H] and Hasse [Ha 3]. However, Ron Evans has pointed out to me that Howard Mitchell [Mi] in 1916 considered Jacobi sums in connection with the number of points of the Fermat curve in arbitrary finite fields, and proved the first Davenport–Hasse relation between Jacobi sums in a finite field and in a finite extension.

Dwork introduced for the first time spaces of power series with p-adic coefficients tending to zero like a geometric series (he denotes such spaces by $L(b)$). He represents Frobenius on factor spaces $L(b)/DL(b)$, where D is an appropriate differential operator, and gets the Gauss sums as eigenvalues in this context, using a simple p-adic trace formula.

Washnitzer–Monsky saw the advantage of taking the union

$$\bigcup_{b>0} L(b),$$

to obtain a more functorial theory. They also introduced in addition certain affine rings over the p-adic numbers, lifting affine rings of hypersurfaces in characteristic p.

Several years ago, Honda looked at Gauss sums again in connection with the Jacobian of the Fermat curve [Ho], and conjectured an expression as limit of certain factorials. This was proved by Katz by taking the action of Frobenius on rather fancy cohomology (unpublished letter to Honda). More recently, Gross–Koblitz recognized this limit as being precisely the value of the p-adic gamma function [Gr–Ko]. In a course (1978), Katz showed that by using the Washnitzer–Monsky spaces and other techniques of Dwork (e.g. his special power series, and estimates of growths of certain coefficients), he could give an elementary proof that the eigenvalue of Frobenius acting on the space associated with the curve

$$y^p - y = x^N$$

was equal to the appropriate expression involving the p-adic gamma function. To establish the equality of this eigenvalue with the Gauss sum, he still referred to the rather extensive theory of Washnitzer–Monsky and their "trace formula," without going through the rather elaborate proofs.

On the other hand, Dwork in [Bo] showed that he could recover the Gross–Koblitz formula by working entirely with his spaces, and by using the completely elementary trace formula which he had proved already in his first paper on the zeta function of a hypersurface [Dw 1]. This paper, and a subsequent one, contained a concrete instance of what Serre [Se 3] recognized as p-adic Fredholm theory, giving a self-contained treatment which systematized Dwork's proofs in those respects which could be viewed as abstract nonsense.

Thus finally it was possible to give a completely elementary proof of the Gross–Koblitz formula. The exposition given here follows Katz in first obtaining the eigenvalue of Frobenius in terms of the gamma function. The second part, getting the Gauss sums, was worked out in collaboration with Dwork, mixing in what seemed the simplest way his spaces and those of Washnitzer–Monsky.

It also turned out that the use of the Artin–Schreier curve was unnecessary for the derivation of the formula, so the connection with that curve is postponed to the next chapter. I am much indebted to Katz for a number of illuminating comments. For instance, although for the special Artin-Schreier curve $y^p - y = x^N$ we can work ad hoc, Katz in his thesis [Ka 2] worked out completely in the general case the relations between Dwork cohomology and Washnitzer–Monsky cohomology. My purpose here was to derive the Gross–Koblitz formula essentially as simply as possible, and to avoid such general theories.

§1. The Basic Spaces

Let p be an odd prime, N a positive integer prime to p, and $q = p^r$ such that $q \equiv 1 \mod N$. We let

$$0 \leq j \leq N - 1 \quad \text{and} \quad a = j\frac{q-1}{N}.$$

We let

$$\omega_q : \mathbf{F}_q^* \to \boldsymbol{\mu}_{q-1}$$

be the Teichmuller character, and we let

$$\psi_{\pi, q} = \psi_\pi \circ \mathrm{Tr}$$

be the additive character formed with the absolute trace Tr and the character ψ_π defined at the end of the last chapter. We let

$$\tau(\omega_q^a, \psi_{\pi, q}) = -S(\omega_q^a, \psi_{\pi, q})$$

be the negative of the Gauss sum, where

$$S(\chi, \psi) = \sum_{x \in \mathbf{F}_q} \chi(x)\psi(x).$$

We let R be a discrete valuation ring of characteristic 0, complete, such that the maximal ideal contains the prime p and also contains π. We let K be the quotient field of R.

We shall work with the ring of power series $R[[x]]_K$ whose coefficients have bounded denominators. This is also given by

$$R[[x]]_K = R[[x]][1/p].$$

We want a vector space

$$\mathscr{A}_{j, \pi} \subset R[[x]]_K$$

such that, if we put

$$H_{j, \pi} = \mathscr{A}_{j, \pi} \frac{dx}{x} \bigg/ d\mathscr{A}_{j, \pi},$$

then dim $H_{j, \pi} = 1$ and $H_{j, \pi}$ is a representation space for the Frobenius map Φ_q^* having $\tau(\omega_q^a, \psi_{\pi, q})$ as eigenvalue. We shall construct such a space.

We shall use an **embedding**

$$\mathrm{emb}_j : R[[x]] \to R[[x]]$$

given by

$$\varphi(x) \mapsto x^j \exp(-\pi x^N)\varphi(x^N).$$

We have a corresponding embedding into differential forms

$$\mathrm{emb}_j^* : R[[x]] \to R[[x]] \, dx$$

given by

$$\varphi(x) \mapsto x^j \exp(-\pi x^N) \varphi(x^N) \frac{dx}{x}.$$

(We use here the assumption that $j \geq 1$ so that no negative powers of x occur in the right-hand side.)

We have a commutative diagram

DW 1.

$$
\begin{array}{ccc}
R[[x]] & \xrightarrow{\ \text{emb}_j\ } & R[[x]] \\
{\scriptstyle ND_j}\Big\downarrow & & \Big\downarrow{\scriptstyle d} \\
R[[x]] & \xrightarrow[\ \text{emb}_j^*\ ]{} & R[[x]]\ dx
\end{array}
$$

where D_j is the **differential operator** given by

$$D_j = x\frac{d}{dx} - \pi x + \frac{j}{N}.$$

This is immediate from the rule giving the derivative of a product.

Remark. For any α not equal to an integer ≤ 0, one verifies at once that the differential operator

$$D = x\frac{d}{dx} - \pi x + \alpha$$

is injective on power series, by looking at its effect on the term of lowest degree.

For each positive number δ we define

$L(\delta) = K$-vector space of power series $\sum a_n x^n$ in $R[[x]]_K$ such that

$$\text{ord } a_n - \delta n \to \infty.$$

This condition could also be written $\text{ord } a_n = \delta n + \infty(n)$.

Remark. This is a variation of Dwork's definition, and one may think of elements of $L(\delta)$ as functions holomorphic on the *closed* disc of "radius" p^δ. As usual, we take

$$\text{ord} = \text{ord}_p.$$

89

We define

$$L(0+) = \bigcup_{\delta} L(\delta).$$

We let

$$\mathcal{A}_{j,\pi} = \text{Image of } L(0+) \text{ under emb}_j$$
$$= \text{space of power series of the form } x^j \exp(-\pi x^N)\varphi(x^N) \text{ with}$$
$$\varphi(x) \in L(0+).$$

This is the space used to define $H_{j,\pi}$ by the formula above. Thus we have an isomorphism

$$L(0+)/D_j L(0+) \approx H_{j,\pi} = \mathcal{A}_{j,\pi} \left. \frac{dx}{x} \right/ d\mathcal{A}_{j,\pi}$$

coming from commutative diagram **DW 1**, and induced by emb$_j$.

Lemma 1.1. *Let $\alpha \in \mathbf{Z}_p$ and suppose that α is not an integer ≤ 0. Let*

$$D = x \frac{d}{dx} - \pi x + \alpha.$$

(i) *We have a direct sum decomposition*

$$L(0+) = K \oplus DL(0+).$$

(ii) *If $\delta \geq 1/(p-1)$, then we have a direct sum decomposition*

$$L(\delta) = K \oplus DL(\delta).$$

Proof. We wish first to write an arbitrary series $\varphi(x)$ as some constant plus Dg for some g. We first solve this problem for powers of x. We have for integers $m \geq 0$:

$$Dx^m = (m + \alpha)x^m - \pi x^{m+1},$$

so that

$$x^{m+1} = \frac{1}{\pi}(m + \alpha)x^m - D\left(\frac{1}{\pi}x^m\right).$$

Recursively, we obtain

$$x^{m+1} = \frac{1}{\pi^{m+1}}(m + \alpha)(m - 1 + \alpha)\cdots(\alpha)$$

$$-D\left[\frac{1}{\pi}x^m + \frac{1}{\pi^2}(m + \alpha)x^{m-1} + \cdots\right.$$

$$\left. + \frac{1}{\pi^{i+1}}(m + \alpha)\cdots(m - i + 1 + \alpha)x^{m-i} + \cdots\right].$$

Hence we have proved what we wanted for powers of x, of degree ≥ 1.

Next, we see what happens for a series in $L(0+)$:

$$(*) \quad \sum_{m=0}^{\infty} b_{m+1}x^{m+1} = \sum_{m=0}^{\infty} b_{m+1}\frac{1}{\pi^{m+1}}(m + \alpha)\cdots(\alpha)$$

$$-D\sum_{n=0}^{\infty} x^n\left[\sum_{m=n}^{\infty} b_{m+1}\frac{1}{\pi\pi^{m-n}}(m + \alpha)\cdots(n + 1 + \alpha)\right].$$

This gives a formal solution, and we want to see that the right-hand side lies in

$$K + DL(0+).$$

For this we need a lemma giving estimates for binomial coefficients.

Lemma 1.2. *Let $\alpha \in \mathbf{Z}_p$, and let n be a positive integer. Then*

$$\text{ord } \alpha(\alpha - 1)\cdots(\alpha - n + 1) \geq -\frac{s(n)}{p - 1} \geq -1 - \log_p n.$$

Proof. Obvious, from

$$\frac{\alpha(\alpha - 1)\cdots(\alpha - n + 1)}{\pi^n} = \frac{n!}{\pi^n}\binom{\alpha}{n},$$

from the fact that the binomial coefficient is p-integral, and from

$$\text{ord } n! = \frac{n - s(n)}{p - 1}, \quad \frac{s(n)}{p - 1} \leq 1 + \log_p n.$$

Since the coefficients b_{m+1} tend to 0 like a geometric progression, it is clear that the first series giving the constant term in the formal solution is actually an element of K. As for the second, there is some positive δ such that

$$\text{ord } b_{m+1} \geq \delta(m + 1) + \infty(m).$$

Hence by Lemma 1.2, for $m \geq n$ we get

(**) $\quad \text{ord } b_{m+1} \dfrac{1}{\pi\pi^{m-n}} (m + \alpha)\cdots(n + 1 + \alpha)$

$$\geq \delta(m + 1) - \frac{1}{p - 1} - \frac{s(m - n)}{p - 1} + \infty(m).$$

This proves that the coefficients of x^n tend to 0 like a geometric series, and hence proves that

$$L(0+) = K + DL(0+).$$

There remains to prove that this sum is direct, in other words that $1 \neq Dg$ for $g \in L(0+)$. This is a special case of the following lemma.

Lemma 1.3. *The equation $Dg = 1$ has a unique solution*

$$g(x) = \sum b_n x^n \in K[[x]],$$

and the coefficients satisfy

$$\text{ord } b_n \leq 1 - \frac{1}{p - 1} + \log_p (n + 1).$$

Proof. Write

$$1 = \left(x\frac{d}{dx} + \alpha - \pi x\right)\left(\sum_{n=0}^{\infty} b_n x^n\right).$$

Then

$$1 + \sum \pi b_n x^{n+1} = \sum (nb_n + \alpha b_n)x^n,$$

so that

$$1 = \alpha b_0 \quad \text{and} \quad \pi b_{n-1} = (n + \alpha)b_n.$$

Hence

$$b_n = \frac{\pi^n}{(n + \alpha) \cdots (\alpha)} = \frac{\pi^n}{(n + 1)! \binom{n + \alpha}{n + 1}}.$$

The same estimate as before, using Lemma 1.2 shows that

$$\operatorname{ord} b_n \leq 1 + \frac{1}{p - 1} + \log_p(n + 1),$$

which proves the lemma.

From the lemma we see that the formal solution of $Dg = 1$ cannot lie in $L(0+)$. This concludes the proof of Lemma 1.1(i).

For the second part we disregard the factorials completely (they are p-integral), and use the more trivial estimate

$$\operatorname{ord} b_{m+1} \geq \delta(m + 1) + \infty(m).$$

Then (**) is now estimated by

$$\operatorname{ord} b_{m+1} \frac{1}{\pi^{m-n+1}} (m + \alpha) \cdots (n + 1 + \alpha) \geq \delta(m + 1) - \frac{m - n}{p - 1} + \infty(m)$$
$$\geq \delta n + \infty(n)$$

because $\delta \geq 1/(p - 1)$. This proves the lemma, because the rest of the proof is identical with what we did before.

Theorem 1.4. *We have isomorphisms for $\delta \geq 1/(p - 1)$:*

$$L(\delta)/D_j L(\delta) \approx L(0+)/D_j L(0+) \approx H_{j,\pi},$$

and these spaces are 1-dimensional.

Proof. Immediate from Lemma 1.1.

§2. The Frobenius Endomorphism

On the power series ring $K[[x]]$, we have the Frobenius endomorphism

$$\Phi_p : K[[x]] \to K[[x]]$$

such that $x \mapsto x^p$, and similarly $\Phi_q = \Phi_p^r$. Then for $h(x) \in K[[x]]$ we have

$$\Phi_q^*(h(x)\, dx) = h(x^q)\, dx^q = h(x^q) q x^q \frac{dx}{x}.$$

First we determine the mapping corresponding to the Frobenius under the embedding emb_j^*. Thus we **define**

$$B_{j,\,p} = x^{(pj-j')/N} E_\pi(x)\Phi_p,$$

where j' is the integer satisfying

$$1 \leq j' \leq N - 1 \quad \text{and} \quad j' \equiv pj \bmod N.$$

Then we have commutative diagrams

FR 1.

$$
\begin{array}{ccc}
R[[x]] & \xrightarrow{\;\mathrm{emb}_j\;} & R[[x]] \\
{\scriptstyle B_{j,\,p}}\Big\downarrow & & \Big\downarrow{\scriptstyle \Phi_{p'}} \\
R[[x]] & \xrightarrow[\;\mathrm{emb}_j\;]{} & R[[x]]
\end{array}
$$

and

$$
\begin{array}{ccc}
R[[x]] & \xrightarrow{\;\mathrm{emb}_j^*\;} & R[[x]]\, dx \\
{\scriptstyle pB_{j,\,p}}\Big\downarrow & & \Big\downarrow{\scriptstyle \Phi_p^*} \\
R[[x]] & \xrightarrow[\;\mathrm{emb}_j^*\;]{} & R[[x]]\, dx.
\end{array}
$$

The commutativity is done by direct verification, and is immediate, starting with a power series $\varphi(x)$ in the upper left-hand corner, and using the definitions of the various mappings going around one way, and then the other way.

A direct verification also shows that

$$\Phi_p : \mathcal{A}_{j,\,\pi} \to \mathcal{A}_{j',\,\pi}$$

maps $\mathcal{A}_{j,\,\pi}$ into $\mathcal{A}_{j',\,\pi}$, and furthermore

$$\Phi_p^* \circ d = d \circ \Phi_p.$$

Consequently we obtain a homomorphism

$$\Phi_p^* : H_{j,\,\pi} \to H_{j',\,\pi},$$

while the diagram **FR 1** yields the corresponding diagram

FR 2.
$$
\begin{array}{ccc}
L(0+)/D_j\, L(0+) & \xrightarrow{\text{emb}_j^*} & H_{j,\,\pi} \\
{\scriptstyle p\bar{B}_{j,\,p}}\Big\downarrow & & \Big\downarrow{\scriptstyle \Phi_p^*} \\
L(0+)/D_j\, L(0+) & \xrightarrow[\text{emb}_{j'}^*]{} & H_{j',\,\pi}.
\end{array}
$$

For our purposes, we also want to look at a composite Frobenius endomorphism, so we let

$$
B_{j,\,q} = x^a E_{\pi,\,q}(x)\Phi_q.
$$

Then we have the commutative diagram

FR 3.
$$
\begin{array}{ccc}
L(0+)/D_j L(0+) & \xrightarrow{\text{emb}_j^*} & H_{j,\,\pi} \\
{\scriptstyle q\bar{B}_{j,\,q}}\Big\downarrow & & \Big\downarrow{\scriptstyle \Phi_q^*} \\
L(0+)/D_j L(0+) & \xrightarrow[\text{emb}_j^*]{} & H_{j,\,\pi}.
\end{array}
$$

Thus $q\bar{B}_{j,\,q}$ is an endomorphism of $L(0+)/D_j L(0+)$, corresponding to the endomorphism Φ_q^* under the embedding emb_j^*.

Inverse of the Frobenius Endomorphism

We also want to tabulate formulas about the inverse of the Frobenius endomorphism, actually one-sided. We define it on power series by

$$
\Psi_q : \sum a_n x^n \mapsto \sum_{q|n} a_n x^{n/q},
$$

and on differential forms by

$$
\Psi_q^*\left(x^n \frac{dx}{x}\right) =
\begin{cases}
\dfrac{1}{q} x^{n/q} \dfrac{dx}{x} & \text{if } q|n \\[2mm]
0 & \text{otherwise.}
\end{cases}
$$

Then

$$
\Psi_q \circ \Phi_q = \text{id}.
$$

Again we have the commutation rule (trivially verified)

DW 2.
$$
\Psi_q^* \circ d = d \circ \Psi_q.
$$

Let

$$A_j = A_{j,q} = \Psi_q \circ (x^{-a} E_{\pi,q}(x)^{-1}).$$

This means that A_j is the composite of multiplication by $x^{-a} E_{\pi,q}(x)^{-1}$ followed by Ψ_q. Since $0 < a < q$, it follows that A_j maps x-integral power series into x-integral power series, and we have

$$A_{j,q} \circ B_{j,q} = \text{id}.$$

We shall need the commutation rule

FR 4. $$A_j \circ D_j = q D_j \circ A_j.$$

This can be proved either directly, or by the same pattern as for the corresponding fact for B_j, using **DW 2** and **FR 3**, together with the fact that emb_j is injective. This gives rise to the commutative diagrams:

FR 5.
$$
\begin{array}{ccc}
R[[x]] \xrightarrow{\ \text{emb}_j^* \ } R[[x]] \dfrac{dx}{x} & & R[[x]] \xrightarrow{\ \text{emb}_j \ } R[[x]] \\
\Big\downarrow {\scriptstyle A_{j,q}} \qquad \Big\downarrow {\scriptstyle q\Psi_q^*} & \text{and} & \Big\downarrow {\scriptstyle A_{j,q}} \qquad \Big\downarrow {\scriptstyle \Psi_q} \\
R[[x]] \xrightarrow[\ \text{emb}_j^* \ ]{} R[[x]] \dfrac{dx}{x} & & R[[x]] \xrightarrow[\ \text{emb}_j \ ]{} R[[x]].
\end{array}
$$

We now consider the effect of the inverse of the Frobenius endomorphism on the Dwork spaces. We see directly from the definitions that

$$\Psi_q : L(\delta) \to L(q\delta)$$

maps $L(\delta)$ into $L(q\delta)$.

Furthermore, let $g \in L(\delta)$. Then multiplication by g,

$$f \mapsto gf$$

maps $L(\delta)$ into itself.

Let δ' satisfy $\delta' < \delta < q\delta'$. Let $g \in L(\delta')$. Then

$$\Psi_q \circ g : L(\delta) \to L(\delta)$$

maps $L(\delta)$ into itself. Indeed, this map is obtained as a composite map:

$$L(\delta) \xrightarrow{\ \text{inc.} \ } L(\delta') \xrightarrow{\ g \ } L(\delta') \xrightarrow{\ \Psi_q \ } L(q\delta') \xrightarrow{\ \text{inc.} \ } L(\delta).$$

We shall apply this to the case when $g(x) = E_\pi(x)$. Since $p \geq 3$ we can select

$$\frac{1}{p-1} \leq \delta < \frac{p-1}{p},$$

and then select $\delta' < (p-1)/p^{r+1}$ but close to $(p-1)/p^{r+1}$ such that

$$\delta' < \delta < q\delta'.$$

Then

$$E_{\pi,q}(x) \in L(\delta').$$

Lemma 2.1. *Under the above conditions on δ and δ',*

$$A_{j,q} : L(\delta) \to L(\delta)$$

maps $L(\delta)$ into itself, and induces a homomorphism

$$\bar{A}_{j,q} : L(\delta)/D_j L(\delta) \to L(\delta)/D_j L(\delta).$$

Proof. This is a special case of the preceding discussion, except that the negative power x^{-a} occurs inside the operator A_j. However, since $a < q - 1$, we have already seen that Ψ_q annihilates such negative powers, so $A_{j,q}$ maps $L(\delta)$ into itself. The commutation rule **FR 4** shows that $A_{j,q}$ induces a homomorphism on the factor space mod $D_j L(\delta)$, as desired.

By Lemma 1.1 we know that this factor space is one-dimensional. Consequently, the operator $\bar{A}_{j,q}$ has an eigenvalue on this space, which we denote by $\lambda'_{j,q} = \lambda'_j$. By definition, we have the relation

$$A_j(1) = \lambda'_j + D_j h_j$$

where h_j is a uniquely determined power series, which lies in $L(\delta)$. Our object is to determine λ'_j. We shall show that λ'_j is a Gauss sum.

We have a commutative diagram

FR 6.

$$
\begin{array}{ccc}
L(0+)/D_j L(0+) & \longrightarrow & L(\delta)/D_j L(\delta) \\
\Big\downarrow{\scriptstyle \bar{B}_j} & & \Big\uparrow{\scriptstyle \bar{A}_j} \\
L(0+)/D_j L(0+) & \longrightarrow & L(\delta)/D_j L(\delta).
\end{array}
$$

97

The horizontal maps are the inverses of those obtained from the natural inclusion $L(\delta) \subset L(0+)$, see Theorem 1.4. We know that

$$A_j \circ B_j = \mathrm{id},$$

so we conclude that the eigenvalue of $\bar{B}_j$ is the inverse of the eigenvalue of $\bar{A}_j$.

§3. The Dwork Trace Formula and Gauss Sums

Consider first the complete space $R\langle\!\langle x \rangle\!\rangle_p$, consisting of the power series whose coefficients tend to 0. Each such power series converges on the (closed) unit disc. We may view the powers of x, namely

$$1, x, x^2, \ldots$$

as forming a "basis" for this space. If

$$u: R\langle\!\langle x \rangle\!\rangle_p \to R\langle\!\langle x \rangle\!\rangle_p$$

is an endomorphism, let us work formally, and represent u by an infinite matrix with respect to this basis. We define the **trace** to be the sum of the diagonal elements. For this to make sense, we must have appropriate conditions of convergence, and in §5 we justify the formal computations which we shall make here in light of the p-adic Banach Fredholm theory of completely continuous operators.

Theorem 3.1. *Let* $g(x) \in R\langle\!\langle x \rangle\!\rangle_p$. *Let* $1 \leq a \leq q - 1$. *Then*

$$(q - 1)\, \mathrm{tr}(\Psi_q \circ x^{-a} g(x)) = \sum_\zeta \zeta^{-a} g(\zeta),$$

where the sum is taken over $\zeta \in \boldsymbol{\mu}_{q-1}$.

Proof. Write

$$h(x) = x^{-a} g(x) = \sum a_n x^n.$$

Then $a_n \to 0$ (p-adically). Let

$$\Psi_q(x^i h(x)) = \sum a_{ij} x^j.$$

Then a_{ij} is the coefficient of x^{qj} in $x^i h(x)$, and so

$$a_{ij} = a_{qj-i}.$$

Hence $a_{ii} = a_{(q-1)i}$, and so

$$(q - 1) \sum a_{ii} = (q - 1) \sum a_{(q-1)i}.$$

But

$$\sum_{\zeta} h(\zeta) = \sum_{\zeta, n} a_n \zeta^n,$$

and in this sum, the n-th term is equal to 0 unless n is divisible by $q - 1$. The remaining terms give precisely the desired value stated in the theorem.

Let

$$\omega_q : F(q)^* \to \mu_{q-1}$$

be the Teichmuller character, such that

$$\omega_q(c) \equiv c \bmod p.$$

The **Gauss sum** is defined by

$$S_q(\chi, \psi) = \sum_{c \in F(q)^*} \chi(c)\psi(c).$$

We apply the preceding discussion with $g(x) = E_{\pi, q}(x)^{-1}$. Then Theorem 3.3 of Chapter 14 yields:

Theorem 3.2. *Let* $a = j(q - 1)/N$ *and* $\psi_{\pi, q} = \psi_\pi \circ T$, *where* **T** *is the absolute trace to the prime field. Then*

$$(q - 1) \operatorname{tr} A_{j,q} = S_q(\omega_q^{-a}, \psi_{\pi, q}^{-1}).$$

Note that the additive character is determined in the usual canonical way (composing with the trace) from the character ψ_π on the prime field. Hence we shall usually omit the additive character from the notation.

Remark. In the next theorem, we assume that the formalism of determinants works in the present situation. This is justified in the last section of the chapter. Specifically, the operator $A_{j,q}$ is viewed as an endomorphism of $L(\delta)$. A Banach basis consists of appropriate scalar multiples of the powers x^n, and the argument used in Theorem 3.1 applies to this case to yield Theorem 3.2. Cf. §5, Example from Dwork Theory.

In the next theorem, we follow usual notation, and let

$$\tau_q = -S_q$$

be the negative of the Gauss sum.

Theorem 3.3. *Let* $1 \leq j \leq N - 1$, *and* $a = j(q - 1)/N$, *where* N *divides* $q - 1$. *Let* δ *be a rational number such that*

$$\frac{1}{p - 1} \leq \delta < \frac{p - 1}{p}.$$

Let $W = L(\delta)/D_j L(\delta)$. *Let* λ'_j *be the eigenvalue of* $\bar{A}_j$ *on* W. *Then*

$$\lambda'_j = \tau_q(\omega_q^{-a}, \psi_{\pi, q}^{-1}).$$

Proof. By **FR 4** of §2, we have a commutative diagram with exact rows:

$$
\begin{array}{ccccccccc}
0 & \longrightarrow & L(\delta) & \overset{D_j}{\longrightarrow} & L(\delta) & \longrightarrow & W & \longrightarrow & 0 \\
 & & {\scriptstyle qA_j}\downarrow & & {\scriptstyle A_j}\downarrow & & {\scriptstyle \bar{A}_j}\downarrow & & \\
0 & \longrightarrow & L(\delta) & \underset{D_j}{\longrightarrow} & L(\delta) & \longrightarrow & W & \longrightarrow & 0.
\end{array}
$$

By the additivity of the trace (cf. Proposition 5.6 below), we have

$$\operatorname{tr} A_j = \operatorname{tr} qA_j + \operatorname{tr} \bar{A}_j.$$

Since W is 1-dimensional, $\operatorname{tr} \bar{A}_j$ is the eigenvalue of $\bar{A}_j$. The theorem is now a direct consequence of Theorem 3.2.

§4. Eigenvalues of the Frobenius Endomorphism and the p-adic Gamma Function

By Lemma 1.1(i) we know that

$$\omega_j = x^j \exp(-\pi x^N) \frac{dx}{x}$$

represents a basis of $H_{j, \pi}$. Consequently there is some element $\lambda(j, \pi, N)$ such that

(1) $$\Phi_p^*(\omega_j) = \lambda(j, \pi, N)\omega_{j'}, \quad \text{in } H_{j', \pi}.$$

We shall relate this element λ to the p-adic gamma function.

Note that Φ_q^* is an endomorphism of $H_{j, \pi}$, so its eigenvalue λ_q is given by

(2) $$\lambda_q = \lambda(j, \pi, N)\lambda(j', \pi, N) \cdots \lambda(j^{(r-1)}, \pi, N).$$

Theorem 4.1. *We have*

$$\lambda(j, \pi, N) = \frac{-p\Gamma_p\left(1 - \dfrac{j'}{N}\right)}{\pi^{(pj - j')/N}}.$$

Proof. (Following Katz.) We have

$$\Phi_p^*\left(x^j \exp(-\pi x^N)\,\frac{dx}{x}\right) = \lambda x^{j'} \exp(-\pi x^N)\,\frac{dx}{x} + dg$$

with some $g \in R[[x]]_K$, that is g is a power series with bounded denominators. Expanding out, we find

$$p \sum \frac{(-\pi)^n}{n!} x^{npN + pj}\,\frac{dx}{x} = \lambda(j, \pi, N) \sum \frac{(-\pi)^m}{m!} x^{mN + j'}\,\frac{dx}{x} + dg.$$

Equating coefficients, using $npN + pj = mN + j'$, we obtain

(3) $$p\,\frac{(-\pi)^n}{n!} = \lambda(j, \pi, N)\,\frac{(-\pi)^m}{m!} + O(npN + pj).$$

Note that $O(npN + pj) = O(n + j/N)$, and

$$m = np + \frac{pj - j'}{N}.$$

This yields

$$p = \lambda(j, \pi, N)(-\pi)^{m-n}\,\frac{n!}{m!} + O\!\left(n + \frac{j}{N}\right)\frac{n!}{\pi^n}.$$

On the other hand, we have

(4) $$\Gamma(1 + m) = (-1)^{1+m}\,\frac{m!}{n!\,p^n}.$$

This is immediate from the definition of the p-adic gamma function, and the fact that a positive integer $i \leq m$ divisible by p can be written in the form

$$i = pi_0 \quad \text{with} \quad i_0 \leq n + \frac{pj - j'}{pN},$$

and therefore $i_0 \leq n$. Hence the denominator $n!\,p^n$ is exactly the product of such i divisible by p.

Substituting (4) in (3), using the fact that $\pi^{p-1} = -p$, and therefore $\pi^{n(p-1)} = (-p)^n$, we find

(5)
$$-p = \frac{\lambda(j, N, \pi)}{\Gamma(1 + m)} \pi^{(pj - j')/N} + O\left(n + \frac{j}{N}\right)\frac{n!}{\pi^n}.$$

The following lemma then concludes the proof.

Lemma. *There exists a sequence of positive integers n such that*

(i) $n \to -j/N$ *p-adically, and hence* $m \underset{p}{\to} -j'/N$;
(ii) $\operatorname{ord}(n + j/N) + \operatorname{ord}(n!/\pi^n) \to \infty.$

Proof. First we simplify slightly the exponent in (ii). We have:

$$\operatorname{ord}\left(n + \frac{j}{N}\right) + \operatorname{ord}\frac{n!}{\pi^n} = \operatorname{ord}\left(n + \frac{j}{N}\right) + \frac{n - s(n)}{p - 1} - \frac{n}{p - 1}$$

$$= \operatorname{ord}\left(n + \frac{j}{N}\right) - \frac{s(n)}{p - 1}.$$

Let $q = p^r$ be such that N divides $q - 1$. Write the expansion

$$-\frac{j}{N} = \frac{j(q - 1)}{N(1 - q)} = \frac{j(q - 1)}{N}(1 + q + q^2 + \cdots).$$

There is also a finite expansion

$$\frac{j(q - 1)}{N} = a_0 + a_1 p + \cdots + a_{r-1}p^{r-1}$$

with standard integral representatives $0 \le a_i \le p - 1$. Let

$$n = \frac{j(q - 1)}{N}(1 + q + \cdots + q^{M-1}).$$

For M tending to infinity, we see that n approaches $-j/N$ to satisfy (i). Furthermore

$$\operatorname{ord}\left(n + \frac{j}{N}\right) \ge M \cdot \operatorname{ord} q = Mr.$$

On the other hand, since $j(q - 1)/N < q - 1$, we have

$$s(n) \le M((p - 1)r - 1).$$

The second condition of the lemma is then obviously satisfied also. This concludes the proof of the lemma, and of Theorem 4.1.

Theorem 4.2. *Let λ_q be the eigenvalue of Frobenius Φ_q^* on $H_{j,\pi}$. Then*

$$\lambda_q = \tau_q(\omega_q^a, \psi_{\pi,q}).$$

Proof. This follows from Theorem 3.3, the remark at the end of §2, **FR 3**, and the fact that

$$\tau(\chi, \psi)\tau(\chi^{-1}, \psi^{-1}) = q.$$

Let a be a positive integer, with $1 \le a < q - 1$ and $q = p^r$. Write

$$a = a_0 + a_1 p + \cdots a_{r-1}p^{r-1}$$

with integers a_i such that $0 \le a_i \le p - 1$. As usual, define

$$s(a) = \sum a_i.$$

Theorem 4.3 (Gross–Koblitz formula).

$$\tau_q(\omega_q^a, \psi_{\pi,q}) = (-1)^r q \pi^{-s(a)} \prod_{i=0}^{r-1} \Gamma_p\left(1 - \left\langle \frac{p^i a}{q-1} \right\rangle\right).$$

Proof. We merely put together Theorem 4.2, and the expression for the eigenvalue obtained in Theorem 4.1. The only quantity remaining to be worked out is the power of π appearing on the right-hand side, and this follows from the following lemma.

Lemma 4.4. *Let $a = j(q-1)/N$. Then*

$$\sum_{i \bmod r} \frac{pj^{(i)} - j^{(i+1)}}{N} = s\left(j \frac{q-1}{N}\right) = s(a).$$

Proof. This is essentially the easy Lemma 1 of Chapter 1, §2. Indeed,

$$\frac{j^{(i)}}{N} = \frac{a^{(i)}}{q-1}$$

where $a^{(i)}$ is defined by

$$1 \le a^{(i)} < q - 1 \quad \text{and} \quad \frac{a^{(i)}}{q-1} = \left\langle \frac{p^i a}{q-1} \right\rangle.$$

With this change of variables from j to the corresponding a, the present lemma is identical with Lemma 1, loc. cit.

Let $\mathfrak{P}$ be the prime in the algebraic closure of $\mathbf{Q}_p$. In the next theorem we show how Stickelberger's result on the Gauss sum mod $\mathfrak{P}$ follows from the Gross–Koblitz formula.

Theorem 4.5 (Stickelberger). *Let* $\gamma(a) = \prod a_i!$. *Then*

$$\pi^{-s(a)}\tau(\omega_q^{-a}, \psi_{\pi, q}) \equiv \frac{1}{\gamma(a)} \mod \mathfrak{P}.$$

Proof. From the formula $\tau\bar{\tau} = q$, and Theorem 4.3 we obtain

(*)
$$\frac{\tau(\omega_q^{-a}, \psi_{\pi,q}^{-1})}{\pi^{s(a)}} = \frac{(-1)^r}{G(a)},$$

where

$$G(a) = \prod_i \Gamma_p\left(1 - \left\langle \frac{p^i a}{q - 1} \right\rangle\right) = \prod_i \Gamma_p\left(1 - \left\langle \frac{p^{r-i}a}{q - 1} \right\rangle\right).$$

Replacing π by $-\pi$ in the left hand side of (*) yields

$$\pi^{-s(a)}\Gamma(\omega_q^{-a}, \psi_{\pi, q})(-1)^{s(a)}.$$

The theorem then follows from the next lemma.

Lemma 4.6.

$$\Gamma_p\left(1 - \left\langle \frac{p^{r-i}a}{q - 1} \right\rangle\right) \equiv (-1)^{1 + a_i}a_i! \mod p.$$

Proof. Note that

$$\Gamma_p(1 + a_i) = (-1)^{1 + a_i}a_i!$$

This is true by definition if $a_i \geq 2$ because $a_i \leq p - 1$. It is also true when $a_i = 1$ and $a_i = 0$ by the direct computations of Chapter 14, §1. On the other hand, by Lemma 1.1 of Chapter 14, we know that

$$\Gamma_p(1 + x) \equiv \Gamma_p(1 + y) \quad \text{if} \quad x \equiv y \mod p.$$

Thus it suffices to prove that

$$-\left\langle \frac{p^{r-i}a}{q-1}\right\rangle \equiv a_i \bmod p.$$

As usual, we write cyclically

$$a = a_0 + a_1 p + \cdots + a_{r-1} p^{r-1}$$
$$pa \equiv a_{r-1} + a_0 p + \cdots + a_{r-2} p^{r-1} \bmod q - 1$$
$$\vdots$$

The desired congruence follows at once, thus proving the lemma and the theorem.

§5. *p*-adic Banach Spaces

The purpose of this section is to do as rapidly as possible the linear algebra in *p*-adic Banach spaces justifying the use of determinants and traces in this chapter. Thus we cover only part of Serre's paper [Se 3], whose exposition leaves nothing to be desired.

Let R be a complete discrete valuation ring, with quotient field K, maximal ideal m, residue class field k, and prime element π.

By a **Banach space** E over K, we mean a normed space which is complete, and such that the norm $x \mapsto |x|$ satisfies

$$|cx| = |c\|x| \quad \text{and} \quad |x + y| \leq \max\{|x|, |y|\}$$

for $c \in K$ and $x, y \in E$. We also assume that the value group $|x|$ (for $x \in E$, $x \neq 0$) is the same group of positive real numbers as the value group of K^*. So for each $x \in E$, there exists $c \in K$ such that $|x| = |c|$.

Let E, F be Banach spaces over K. Let $L(E, F)$ be the vector space of continuous linear maps

$$u: E \to F.$$

As usual, we can define a **norm** on $L(E, F)$ by

$$|u| = \sup_{x \neq 0} \frac{|ux|}{|x|}.$$

Also as usual, we have

$$|u| = \sup_{|x| \leq 1} |ux|.$$

By an **isomorphism**, we mean a norm-preserving continuous linear map having an inverse.

Example. Let I be a set of indices, and let $C(I, K)$ be the set of families $x = (x_i)_{i \in I}$, $x_i \in K$, such that x_i tends to 0 as $i \to \infty$. By this we mean that given ε there exists a finite set of indices S such that for $i \notin S$, we have $|x_i| < \varepsilon$. The reader may as well think of the positive integers where $i = 1, 2, 3, \ldots$. We define

$$|x| = \sup |x_i|.$$

Then $C(I, K)$ is a Banach space.

More generally, if F is a Banach space, we may form families (x_i) with $x_i \in F$ such that $x_i \to 0$. Such families with the sup norm form a Banach space $C(I, F)$.

A family $\{e_i\}$ in a Banach space E will be called a **Banach basis** if every element $x \in E$ can be written uniquely as a series

$$x = \sum_{i \in I} x_i e_i \quad \text{with } x_i \to 0$$

and $|x| = \sup |x_i|$. It is clear that the space $C(I, K)$ has such a basis, with $e_i(i) = 1$ and $e_i(i') = 0$ if $i' \neq i$.

Let E be a Banach space and let E_0 denote the subset of elements $x \in E$ such that $|x| \leq 1$. If $\{e_i\}$ is a Banach basis, then E_0 consists of the set of all elements

$$\sum x_i e_i$$

such that $|x_i| \leq 1$ for all i.

Lemma 5.1. *Let* $\mathfrak{m}$ *be the maximal ideal of* R. *Let* $\bar{E} = E_0/\mathfrak{m}E_0$. *A family* $\{e_i\}$ *in* E *is a Banach basis if and only if all* $e_i \in E_0$, *and their images* $\bar{e}_i$ *in* $\bar{E}$ *form an algebraic basis of* $\bar{E}$ *as vector space over* k.

Proof. Suppose that $\{e_i\}$ is a Banach basis for E. Then first it is clear that every element of $\bar{E}$ can be written as a linear combination of the $\bar{e}_i$ with coefficients in k, and such a combination is unique, as one sees by lifting back to E_0.

Conversely, suppose $\{\bar{e}_i\}$ is an algebraic basis for $\bar{E}$. Any element $x \in E_0$ can be written

$$x = \sum x_i^{(1)} e_i + \pi x^{(1)} \quad \text{with } x^{(1)} \in E_0.$$

Iterating, and expressing $x^{(1)}$ in a similar way, we get an expression

$$x = \sum x_i e_i \quad \text{with } x_i \in R \text{ and } x_i \to 0.$$

Such an expression is unique, for if we have a relation

$$\sum x_i e_i = 0, \quad \text{not all } x_i = 0,$$

then first we may divide by the highest power of π dividing all x_i, so that not all x_i are divisible by π, and then we may reduce mod $\mathfrak{m}$ to get a contradiction. Finally if $|x| = 1$, we have

$$|x| = \sup |x_i|,$$

and the same relation holds for any x by multiplying with an appropriate power of π. This proves the lemma.

Proposition 5.2. *Every Banach space over K is isomorphic with a space $C(I, K)$. Equivalently, every Banach space has a Banach basis.*

Proof. Immediate from the lemma, by lifting an algebraic basis from $\bar{E}$ back to E_0.

Corollary. *Every closed subspace F of a Banach space E has a complementary closed subspace F' such that $E \approx F \times F'$.*

Proof. Let F be a closed subspace of E. By the theorem, there exists a Banach basis $\{e_i'\}$ of the factor space E/F. Let e_i be a representative of e_i' in E such that $|e_i| \leq 1$. The map

$$u : e_i' \mapsto e_i$$

extends to a continuous linear map $E/F \to E$, of norm ≤ 1. Then the map

$$E/F \times F \to E$$

given by

$$(\bar{x}, y) \mapsto u\bar{x} + y$$

is an isomorphism, as is immediately verified. This proves the corollary.

Let F be a Banach space. Let I be a set of indices, and let $B(I, F)$ be the set of bounded maps of I into F, i.e., the set of bounded families $(f_i)_{i \xi I}$ with the sup norm. Then $B(I, F)$ is a Banach space.

Proposition 5.3. *Let $E = C(I, K)$ and let $\{e_i\}$ be a Banach basis. Then we have an isomorphism*

$$L(E, F) \to B(I, F)$$

given by

$$u \mapsto (ue_i)_{i \in I}.$$

Proof. Let φ be the map from $L(E, F)$ to $B(I, F)$ as given in the statement of the proposition. Conversely, if (f_i) is a bounded family in F, define a map

$$\psi : B(I, F) \to L(E, F)$$

by associating to (f_i) the map u such that

$$u\left(\sum x_i e_i\right) = \sum x_i f_i.$$

It is then immediate that φ, ψ are continuous linear, inverse to each other, and have norms ≤ 1, so that they are isomorphisms. This proves the proposition.

The proposition will be used especially when F has a Banach basis, so that $F = C(J, K)$ for some set of indices J. Each element ue_i can then be written uniquely

$$ue_i = \sum_j c_{ij} e'_j$$

where (e'_j) is the natural Banach basis for $C(J, K)$. Thus u has an associated matrix (c_{ij}), relative to the bases (e_i) and (e'_j). By the definitions, and Proposition 5.3, we have

$$|u| = \sup_{i,j} |c_{ij}|.$$

Example. Let $F = K$ so that $L(E, K)$ is the dual space, denoted by E^*. If we let $ue_i = c_i$, then

$$|u| = \sup_i |c_i|.$$

Let $u \in L(E, F)$. We say that u is **completely continuous** if u is a limit of elements of $L(E, F)$ with finite dimensional image. (The limit of course taken with respect to the norm on $L(E, F)$.) We denote by $CC(E, F)$ the space of such completely continuous linear maps. It is obviously a Banach subspace of $L(E, F)$.

Let E, E', E'' be Banach spaces, and let

$$E \xrightarrow{u} E' \xrightarrow{v} E''$$

be continuous linear maps. If u or v is completely continuous, then so is $v \circ u$. This is immediate from the definition. In particular, $CC(E, E)$ is a two-sided ideal in $L(E, E)$.

Suppose now that $F = C(J, K)$ with Banach basis (f_j). Let $u \in L(E, F)$. For each $x \in E$ we can write

$$ux = \sum u_j(x) f_j$$

where $u_j \in E^*$ is an element of the dual of E, i.e. a functional. We define

$$r_j(u) = |u_j|.$$

Then

$$|u| = \sup_j r_j(u) = \sup_j |u_j|,$$

directly from the definitions.

Let $E = C(I, K)$, and let (c_{ij}) be the matrix associated with u, relative to the bases (e_i) and (f_j). Then

$$|u_j| = \sup_i |c_{ij}|.$$

This is clear from the example following Proposition 5.3.

Proposition 5.4. *Let $F = C(J, K)$, with Banach basis (f_j). The map*

$$u \mapsto (u_j)$$

gives an isomorphism

$$CC(E, F) \approx C(J, E^*)$$

between $CC(E, F)$ and the space of families (u_j) of elements in E^* such that $u_j \to 0$. In particular, an element $u \in L(E, F)$ is completely continuous if and only if $u_j \to 0$.

Proof. First suppose u has finite dimensional image, so without loss of generality, we may assume the image is one-dimensional. Then $u(x) = v(x) f$ for some $f \in F$, and v is a functional. Then $u_j = f_j v$, where f_j is the j-th coordinate of f, so clearly $u_j \to 0$. Next let $u^{(n)} \in L(E, F)$ approach u, and have finite dimensional images. Then

$$u_j^{(n)} \to u_j$$

uniformly, so $u_j \to 0$ also in this general case.

Finally, let (v_j) be a family of elements in the dual E^* such that $v_j \to 0$. Define u by

$$ux = \sum v_j(x) f_j.$$

Then u is certainly continuous linear. The partial sums u_S defined for every finite set S of indices by

$$u_S(x) = \sum_{j \in S} v_j(x) f_j$$

approach u in the norm of $L(E, F)$, and have finite rank. Hence the correspondence between $CC(E, F)$ and $C(J, E^*)$ is bijective. It is immediately verified to be norm preserving. This concludes the proof.

Example from Dwork theory. Let $E = L(\delta)$, with the norm defined as follows. If

$$f(x) = \sum a_n x^n$$

is in $L(\delta)$, we define

$$|f|_\delta = \sup \left| \frac{a_n}{p^{n\delta}} \right|.$$

Then E is a Banach space. We assume that δ is rational, and that the constant field K has been extended by a finite extension if necessary so that p^δ lies in K. The powers

$$(p^\delta x)^n, \qquad n = 0, 1, 2, \ldots$$

form a Banach basis for E. If $\delta' < \delta$, then the inclusion

$$L(\delta) \to L(\delta')$$

is completely continuous. This is obvious by looking at the associated matrix, and using Proposition 5.4.

On the other hand, if $g \in L(\delta')$, then the map $f \mapsto fg$ is a continuous linear map of $L(\delta')$ into itself. Furthermore, the Dwork map Ψ_q is a continuous linear map of $L(\delta')$ into $L(q\delta')$. It follows that the Dwork operator $A_{j,q}$ is completely continuous on the space $L(\delta)$.

We now come to the results on Fredholm determinants which were used in the context of Dwork theory. First we make some remarks about determinants in finite dimensional spaces.

Let L be a free module over a commutative ring, and let u be an endomorphism of L such that $u(L)$ is contained in a finitely generated submodule. Let M be a submodule of L which is finitely generated, free, and contains $u(L)$. Let

$$u_M : M \to M$$

be the restriction of u to M. The polynomial

$$\det(I - tu_M)$$

is well defined, and it is easy to verify that it does not depend on the choice of M. We may therefore denote it by

$$\det(I - tu) = 1 + c_1 t + \cdots + c_m t^m + \cdots .$$

The coefficients c_m can be expressed in terms of u as follows.

Let (e_i) be a basis of L. Let S be a finite subset of the set of indices I, with card $S = m$. Let $A = (a_{ij})$ be the matrix of u with respect to this basis, and let

$$\det_S(A) = \text{determinant of the submatrix of } a_{ij} \text{ with } i, j \in S.$$

Then

$$\boxed{c_m = (-1)^m \sum_{|S| = m} \det_S(A),}$$

where the sum is taken over subsets S having m elements.

This formula actually relates to a finite square matrix. It suffices to prove it when the matrix consists of algebraically independent elements in characteristic 0 (by specializing afterward). In that case, the matrix has distinct eigenvalues, and the formula is obvious if the matrix is diagonal. To get the formula in general, all we need is an invariant characterization of the terms in the sum over S. This is provided by the second expression for the coefficients:

$$c_m = (-1)^m \text{ tr} \bigwedge^m u,$$

where $\bigwedge^m u$ is the m-th exterior power of u, acting on $\bigwedge^m M$, and M may be assumed to be a finite dimensional vector space. But in that case, one finds at once that

$$\operatorname{tr} \bigwedge\nolimits^m u = \sum_{|S|=m} \det_S(A),$$

by looking at the trace with respect to the basis

$$\{e_{i_1} \wedge \cdots \wedge e_{i_m}\}.$$

This proves that the coefficients c_m have the value as stated. In particular, $c_1 = -\operatorname{tr} u$ (sum of the diagonal elements), so

$$\det(I - tu) = 1 - (\operatorname{tr} u)t + \cdots.$$

Now suppose E is a Banach space with Banach basis (e_i), $i \in I$. We wish to define $\det(I - tu)$ for any $u \in CC(E, E)$. Suppose first that $|u| \leq 1$. Let E_0 be as before, the subspace of elements x in E such that $|x| \leq 1$. Then $u(E_0) \subset E_0$. For each positive integer n, the endomorphism u defines an endomorphism

$$u(\pi^n) : E_0/\pi^n E_0 \to E_0/\pi^n E_0.$$

We denote $E_0/\pi^n E_0 = E_0(\pi^n)$. If (a_{ij}) is the matrix of u with respect to the basis (e_i), then we know that

$$|u_j| = \sup_i |a_{ij}| \to 0.$$

Hence there exists a finite subset S of indices j such that for all i and all $j \notin S$, the components a_{ij} lie in the ideal (π^n). Hence the polynomial

$$\det(I - tu(\pi^n))$$

is defined, as a polynomial in $R/\pi^n R$. As n tends to infinity, these polynomials form a projective system, and we define

$$\det(I - tu)$$

to be their limit. It is a formal power series, in $R[[t]]$.

If u is arbitrary in $CC(E, E)$, we can find an element $c \in R$, $c \neq 0$, such that $|cu| \leq 1$. Then $\det(I - tcu)$ is defined, as a power series $D(t)$, and we define $\det(I - tu) = D(tc^{-1})$. This is independent of the choice of c.

Proposition 5.5. *Let u be a completely continuous endomorphism of the Banach space E. Let* $(e_i)_{i \in I}$ *be a Banach basis, and let* $A = (a_{ij})$ *be the matrix of u with respect to this basis. Let*

$$\det(I - tu) = \sum_{m=0}^{\infty} c_m t^m.$$

Then:

(i)
$$c_m = (-1)^m \sum_S \det_S(A).$$

    where S ranges over subsets of I with m elements.

(ii) *Given a positive number r, we have* $|c_m| \ll r^m$ *for all m sufficiently large.*

(iii) *If* $\{u_n\}$ *is a sequence in CC(E, E) and* $u_n \to u$, *then*

$$\det(I - tu_n) \to \det(I - tu),$$

    where the convergence is the simple convergence of coefficients.

(iv) *If u has finite rank, then* $\det(I - tu)$ *is the polynomial defined from elementary linear algebra.*

Proof. Suppose first that $|u| \leq 1$. Then the formula for c_m is valid for each $u(\pi^n)$, and so remains valid in the limit. The general case follows by using $c \neq 0$ such that $|cu| \leq 1$.

For (ii), let S be a finite subset of m elements. Each product in the expansion for the determinant $\det_S(A)$ contains m elements, indexed by m distinct indices $j_1, \ldots, j_m$. Let

$$r_j = |u_j| \quad \text{so} \quad r_j \to 0.$$

Then any product of m elements as above is bounded in absolute value by the product $r_1 \cdots r_m$. Hence

$$|\det_S(A)| \leq r_1 \cdots r_m \quad \text{for each } S,$$

so

$$|c_m| \leq r_1 \cdots r_m.$$

Since $r_j \to 0$ it follows that $r_j \leq r$ for all j sufficiently large, and (ii) follows. The third assertion follows trivially from (i).

Finally, suppose u has finite dimensional image. After replacing u by a scalar multiple, we may assume that $|u| \leq 1$. Let F be a finite dimensional subspace of E containing $u(E)$. Let

$$F_0 = F \cap E_0 \quad \text{and} \quad F(\pi^n) = F_0/\pi^n F_0.$$

Then F_0 is a direct summand of E_0, and hence $F_0(\pi^n)$ is a direct summand of $E_0(\pi^n)$, and is a free submodule. Hence the determinant

$$\det(I - tu(\pi^n))$$

can be computed in $F_0(\pi^n)$. Since

$$\det(I - tu) = \lim \det(I - tu(\pi^n)),$$

it follows that

$$\det(I - tu) = \det(I - tu_F),$$

where u_F is the restriction of u to F. Then (iv) follows, and the proposition is proved.

By the estimate of (ii) in the proposition, we conclude that the power series $\det(I - tu)$ represents an entire function of the variable t for values in K, or in any complete extension of K. In particular, $\det(I + u)$ is defined for all $u \in CC(E, E)$.

Corollary 1. *Let* $u, v \in CC(E, E)$. *Then*

$$\det((I - tu)(I - tv)) = \det(I - tu)\det(I - tv).$$

Proof. After multiplying u, v by appropriate scalars, we may assume without loss of generality that $|u| \leq 1$ and $|v| \leq 1$. In that case, we view u, v as acting on E_0, and reduce mod π^n, in which case the formula is true when u, v are replaced by $u(\pi^n)$ and $v(\pi^n)$ respectively. Taking the limit yields the corollary.

Corollary 2. *Let* $u \in CC(E, F)$ *and* $v \in L(F, E)$. *Then*

$$\det(I - tu \circ v) = \det(I - tv \circ u).$$

Proof. By (iii) and (iv) we may assume that u, v have finite rank, in which case the assertion is standard by the linear algebra of finite dimensional spaces.

We define the **trace** tr(u) of an element $u \in CC(E, E)$ as the coefficient of $-t$ in det($I - tu$). If (a_{ij}) is the matrix associated with u with respect to a Banach basis, then as usual,

$$\text{tr}(u) = \sum a_{ii}.$$

This formula shows that $|\text{tr}(u)| \leq |u|$.

Corollary 3. *Assume that K has characteristic 0. Then*

$$\det(I - tu) = \exp\left(-\sum_{m=1}^{\infty} \text{tr}(u^m)\, \frac{t^m}{m!}\right).$$

Proof. The assertion is true if u has finite rank by ordinary linear algebra. Let $\{u_n\}$ be a sequence of such endomorphisms converging to u. Then both the right-hand side and left-hand side with u replaced by u_n converge to the corresponding expressions of the corollary, thus establishing their equality.

Note. The above three corollaries are included to show how to apply Proposition 5.5. They were not needed in the applications of the preceding sections. Only the following proposition was needed.

Proposition 5.6. *Let*

$$0 \to E' \to E \to E'' \to 0$$

be an exact sequence of Banach spaces, and let u', u, u'' be continuous linear maps making the diagram commutative:

$$
\begin{array}{ccccccccc}
0 & \longrightarrow & E' & \longrightarrow & E & \longrightarrow & E'' & \longrightarrow & 0 \\
 & & \downarrow{\scriptstyle u'} & & \downarrow{\scriptstyle u} & & \downarrow{\scriptstyle u''} & & \\
0 & \longrightarrow & E' & \longrightarrow & E & \longrightarrow & E'' & \longrightarrow & 0
\end{array}
$$

If u is completely continuous, so are u' and u'', and we have

$$\det(I - tu) = \det(I - tu')\det(I - tu''),$$

$$\text{tr } u = \text{tr } u' + \text{tr } u''.$$

Proof. We may view E' as a subspace of E, and u' as the restriction of u to E'. Thus u'' is the map induced on E'' as factor space E/E'. By the Corollary of Proposition 5.2 there exists a Banach basis $\{e'_i\}$ for E' and a Banach basis $\{e''_j\}$ for a complementary subspace such that the images of $\{e''_j\}$ in E/E' form a Banach basis. Then $\{e'_i, e''_j\}$ forms a Banach basis for E. From this it is clear

115

that if u is completely continuous, so are u' and u''. Reducing mod π^n yields the desired identity for $u(\pi^n)$, $u'(\pi^n)$, and $u''(\pi^n)$, whence the identity as in the proposition for u, u', u''. The identity for the trace follows from that of the determinant since

$$\det(1 - tu) = 1 - (\operatorname{tr} u)t + \cdots.$$

This concludes the list of properties of the determinant which has been used in the Dwork theory.

Gauss Sums and the Artin–Schreier Curve

16

In this chapter we establish the connection between the spaces used to represent the Frobenius endomorphism, and eigenspaces for a Galois group of automorphisms of the Artin–Schreier curve. This connects Dwork theory, the Washnitzer–Monsky theory, and the fact realized by Monsky that the series $E_\pi(x)$ can be used to construct explicitly such eigenspaces. The first section lays the foundations for the special type of ring under consideration. After that we study the Artin–Schreier equation and the Frobenius endomorphism.

§1. Power Series with Growth Conditions

We let:

R = discrete valuation ring of characteristic 0, complete, with prime element π dividing the prime number p.
K = quotient field of R.
$k = R/\pi$ = residue class field, which has characteristic p.

$R\langle\!\langle x \rangle\!\rangle$ = set of power series

$$(1) \qquad \varphi(x) = \sum_{n=0}^{\infty} a_n x^n \quad \text{with } a_n \in R$$

such that there exists $\delta > 0$ for which

$$\text{ord}_\pi a_n \geq \delta n \quad \text{for all } n \text{ sufficiently large.}$$

It is clear that $R\langle\!\langle x\rangle\!\rangle$ is a ring, which will be called the **Washnitzer–Monsky ring**. Furthermore, the elements of $R\langle\!\langle x\rangle\!\rangle$ are precisely the power series which converge (absolutely) on a disc of radius > 1. Of course, the radius depends on the power series, there is no uniformity required. Later, we shall deal with the Dwork spaces, taking the uniformity into account. Indeed, it will also be useful to consider power series such that a finite number of coefficients may lie in K.

Pick δ rational > 0, and let π^δ denote any rational power of π, well defined up to a root of unity in the integral closure of R. If $\varphi(x)$ as above is an element of $R\langle\!\langle x\rangle\!\rangle$, then we may write

$$(2) \qquad\qquad \varphi(x) = \sum b_n(\pi^\delta x)^n$$

with coefficients b_n (possibly algebraic over R) which are π-integral for all but a finite number of n. Conversely, if a power series $\varphi(x)$ in $R[[x]]$ has a representation as in (2) with coefficients b_n whose denominators are bounded, then it is clear that $\varphi(x) \in R\langle\!\langle x\rangle\!\rangle$.

We let $R\langle\!\langle x\rangle\!\rangle_p$ be the p-adic completion of $R\langle\!\langle x\rangle\!\rangle$. Then $R\langle\!\langle x\rangle\!\rangle_p$ is the ring of power series

$$\sum a_n x^n, \qquad a_n \in R,$$

such that $a_n \to 0$, under the sup norm of coefficients. To prove that this is the completion, we observe that this ring $R\langle\!\langle x\rangle\!\rangle_p$ is p-adically complete, and that $R\langle\!\langle x\rangle\!\rangle$ is dense in it. Furthermore, the polynomial ring $R[x]$ is dense, so that

$$R\langle\!\langle x\rangle\!\rangle_p = R[x]_p$$

is also the p-adic completion of $R[x]$.

The ring $R\langle\!\langle x\rangle\!\rangle_p$ still has the advantage that

$$R\langle\!\langle x\rangle\!\rangle_p \bmod \pi = k[x],$$

in other words, its reduction mod π is the polynomial ring.

The following property is immediately verified.

Let $w \in R\langle\!\langle x\rangle\!\rangle$ (respectively $R\langle\!\langle x\rangle\!\rangle_p$) be such that

$$w \equiv 0 \bmod \pi x.$$

Then the geometric series gives an inverse for $1 - w$ in $R\langle\!\langle x\rangle\!\rangle$ (respectively $R\langle\!\langle x\rangle\!\rangle_p$).

Lemma 1.1. *The ring $R\langle\!\langle x\rangle\!\rangle_p$ is integrally closed.*

Proof. An element $\varphi(x)$ of $R\langle\!\langle x \rangle\!\rangle_p$ not divisible by π can be written

$$\varphi(x) = b_0 + \cdots + b_d x^d + \cdots$$

such that b_d is a unit, and $b_n \equiv 0 \mod \pi$ for $n \geq d + 1$. Then the Manin proof of the Weierstrass preparation theorem given for instance in Chapter 5, §2 applies to show that

$$\varphi(x) = P(x)u(x),$$

where $P(x) = x^d + \cdots + c_0 \in R[x]$ is a polynomial, and $u(x)$ is a unit in $R\langle\!\langle x \rangle\!\rangle_p$. (In that reference, the crucial step occurs when we assert that $\tau(f)$ is invertible, which is true in $R\langle\!\langle x \rangle\!\rangle_p$, by the remark made before the lemma.) This implies that the zeros of $P(x)$ all lie in the unit disc in the algebraic closure of the quotient field of R.

If the quotient $\varphi(x)/\psi(x)$ of elements in $R\langle\!\langle x \rangle\!\rangle_p$ is integral over $R\langle\!\langle x \rangle\!\rangle_p$, then we use the Weierstrass preparation theorem to write

$$\frac{\varphi(x)}{\psi(x)} = \pi^r \frac{P(x)}{Q(x)} u(x),$$

where $P(x), Q(x)$ are polynomials, and $u(x)$ is a unit in $R\langle\!\langle x \rangle\!\rangle_p$. Then without loss of generality, we may assume that $u(x) = 1$. Since elements of $R\langle\!\langle x \rangle\!\rangle_p$ converge on the unit disc, if $Q(x)$ does not divide $P(x)$, we may evaluate the right-hand side at a zero of $Q(x)$ which is not a zero of $P(x)$ to make the right-hand side infinity. An elementary criterion for integrality then shows that $\varphi(x)/\psi(x)$ cannot be integral over $R\langle\!\langle x \rangle\!\rangle_p$, which proves the lemma.

Lemma 1.2. *The ring $R\langle\!\langle x \rangle\!\rangle$ is algebraically closed in $R\langle\!\langle x \rangle\!\rangle_p$, and therefore it is integrally closed (in its quotient field).*

Proof. As pointed out to me by Dwork, this lemma can be viewed as a special case of a result in Dwork–Robba [Dw–Ro], Theorem 3.1.6. The proof given below was derived in collaboration with Dwork.

Let $A = R\langle\!\langle x \rangle\!\rangle$ and $A_p = R\langle\!\langle x \rangle\!\rangle_p$. Let $y \in A_p$ be algebraic over A, satisfying a polynomial equation

$$F(y) = 0$$

with coefficients in A. By the x-topology, we mean the topology of formal power series (high powers of x are close to zero). Given $\alpha \in A_p$, if α' is x-close to α, then α' is also p-close to α.

Let y_0 be a polynomial, in $R[x]$, which is x-close to y. Let $\| \ \|$ be the **Gauss norm** on power series (sup norm of the coefficients) extended to the quotient field $K(A)$. Then $F'(y_0) \in A$, and in particular is holomorphic on a disc of radius > 1. Then for y_0 sufficiently close to y, we have

$$\left\| \frac{F(y_0)}{F'(y_0)^2} \right\| < 1.$$

Hence the Newton sequence

$$y_{n+1} = y_n - \frac{F(y_n)}{F'(y_n)}$$

converges in the Gauss norm in the completion of $K(A)$, and in fact to y itself since A_p is complete, if we pick y_0 sufficiently close to y (closer than any other root of F in the completion of $K(A)$). Cf. for instance my paper, "On quasi algebraic closure," *Ann. of Math.* (1952) pp. 373–392; or also my *Algebraic Number Theory*, Chapter II.

We shall now use another norm to see that the sequence converges to a holomorphic function on a disc of radius $\geq 1 + \varepsilon$ for some ε, from which relatively small sets have been deleted.

To avoid introducing a new letter, let us view K as a subfield of $\mathbf{C}_p$. If t is a real number > 0, we let $D(t, 0)$ be the closed disc of radius t around the origin in $\mathbf{C}_p$. Let $\alpha_1, \ldots, \alpha_m$ be elements of this disc, and let $r_1, \ldots, r_m > 0$. We let

$B = B(t; r, \alpha) =$ set obtained by deleting from $D(t, 0)$ the union of the balls $D(r_j, \alpha_j)$ of radius r_j, centered at α_j.

In the sequel, we assume that $t > 1$ and $r_j < 1$,

If $H(x)$ is a rational function with no poles in a set B as above, then from its factorization into linear factors, we see that the norm

$$\|H(x)\|_B = \sup_{x \in B} |H(x)|$$

is defined.

Lemma 1.2.1. *For any rational function H holomorphic on B, we have*

$$\|H\| \leq \|H\|_B.$$

Proof. We factor the rational function into a product of a constant factor and linear factors of type

$$x - a \quad \text{and} \quad \frac{1}{x - a}, \quad \text{with } a \in \mathbf{C}_p.$$

The Gauss norm of $x - a$ is $\max(1, |a|)$. We pick $x \in B$ to be a unit which is not congruent to any $a \bmod \mathfrak{m}_p$. For such x we have

$$|H(x)| \leq \|H\|_B,$$

and the lemma is then obvious.

Lemma 1.2.2. *Given a rational function $H(x)$, with $\|H\| < 1$, there exists a set $B(t; \alpha, r)$ with $t > 1$ and $r_j < 1$, such that*

$$\|H\|_B < 1.$$

Proof. Factor

$$H(x) = c \prod (1 - a_i x)^{m_i} \prod (x - b_j)^{n_j}$$

where $|a_i| < 1$ and $|b_j| \leq 1$. Then $|c| < 1$ because $\|H\| < 1$. Let N be the number of linear factors, counting multiplicities. Let $s > 1$ be such that $s^N c < 1$. It will suffice to find B such that each factor of H has B-norm $\leq s$. We consider factors of four types:

$$1 - ax, \quad (1 - ax)^{-1}, \quad x - b, \quad (x - b)^{-1}$$

with $|a| < 1$ and $|b| \leq 1$. For factors of the first three types, it suffices to select the radius t of B to be $< s$ and sufficiently small > 1. For the factor of last type, namely $(x - b)^{-1}$, it suffices to delete from B a disc of radius $1/t'$, where $t' < t$ and t' is very close to t. This proves the sublemma.

We now return to the proof of Lemma 1.2. We define:

$\text{Hol}(B)$ = completion of the ring of rational functions
having no poles in B, under the B-norm.

Since the Gauss norm is bounded by the B-norm, every Cauchy sequence for the B-norm is a Cauchy sequence for the Gauss norm. Consequently there is a natural injection of the completions

$$\text{Hol}(B) \to K(x)_{\text{Gauss}}.$$

Observe that $A_p = R\langle\!\langle x \rangle\!\rangle_p$ is contained in the Gauss completion.

Given our element $y \in A_p$ algebraic over A, we first select a polynomial $y_0 \in R[x]$ as in the beginning of the proof, so that

$$\left\| \frac{F(y_0)}{F'(y_0)^2} \right\|$$

is small, and in particular is < 1. We let $H(x) = F(y_0)/F'(y_0)^2$ be this rational function. We then select B as in Lemma 1.2.2 so that $\|H\|_B < 1$. Then the Newton sequence also converges to an element of $\text{Hol}(B)$. This implies that y is the power series expansion on the *closed* disc of radius 1 of a holomorphic function on B. It is now a matter of foundations of p-adic analytic functions that the power series for y also represents the analytic function on some disc of radius > 1. This implies that the coefficients of the power series $y = \varphi(x)$ tend to 0 like some geometric series, thereby proving Lemma 1.2.

For a proof of the foundational fact we have just used, see for instance Amice [Am]. Note that the power series for y may converge only on a disc of radius smaller than the radius of the original set B. For instance, the holomorphic function may have a finite number of poles near but outside the circle of radius 1. The power series will converge only on a disc which does not contain these poles. The proof requires the p-adic analogue of the Mittag-Leffler theorem, in lieu of analytic continuation over the complex numbers.

Let $f_0(Y)$ be an irreducible polynomial of degree d, with coefficients in $k[x]$, leading coefficient 1. By a **lifting** of f_0 we mean a polynomial $f(Y)$ in $R\langle\!\langle x \rangle\!\rangle[Y]$ of the same degree, leading coefficient 1, such that

$$f_0 = f \mod \pi.$$

Then f is necessarily irreducible over $R\langle\!\langle x \rangle\!\rangle_p$. Indeed, the coefficients in a factor are integral over $R\langle\!\langle x \rangle\!\rangle$, so in $R\langle\!\langle x \rangle\!\rangle$ by Lemmas 1.1 and 1.2. Such a factor then reduces to a factor of f_0.

Let y_0 be a root of f_0 and let y be a root of f. Let

$$\mathscr{A} = R\langle\!\langle x \rangle\!\rangle[y], \quad \text{and} \quad \mathscr{A}_0 = k[x, y_0].$$

Then there is a unique homomorphism

$$\mathscr{A} \to k[x, y_0] = \mathscr{A}_0$$

reducing $R\langle\!\langle x \rangle\!\rangle$ mod π, and sending y to y_0. This is a standard fact of elementary field theory. Thus the ideal (π) in $R\langle\!\langle x \rangle\!\rangle$ extends uniquely to a prime ideal of $\mathscr{A}$, and

$$\mathscr{A}_0 = \mathscr{A} \mod \pi.$$

We view f as defining an affine curve V and f_0 as defining its reduction mod π. Thus we shall write

$$\mathscr{A} = \mathscr{A}(V) \quad \text{and} \quad \mathscr{A}/\pi = \mathscr{A}_0(V_0).$$

Then $\mathscr{A}_0(V_0)$ is the ordinary affine ring of V_0 over the field $k = R/\pi$, but $\mathscr{A} = \mathscr{A}(V)$ is a more complicated ring, arising from the work of Washnitzer and Monsky, following Dwork. We shall also say that V is a **lifting** of V_0, corresponding to the lifting f of f_0.

Lemma 1.3. *Let $w \in \mathscr{A}$ and assume $w \equiv 0$ mod π. Then the geometric series*

$$1 + w + w^2 + \cdots$$

converges to an inverse of $1 - w$ in $\mathscr{A}$.

Proof. For convenience, assume that the lifted polynomial $f(Y)$ has coefficients in $R[x]$. This is all that we need in the applications, and the proof is slightly easier in this case. We write

$$w = \pi \sum_{i=0}^{d-1} g_i(x) y^i$$

with $g_i(x) \in R \langle\!\langle x \rangle\!\rangle$. Then for some $\delta > 0$ we also have

$$w = \sum_{i=0}^{d-1} h_i(\pi^\delta x)(\pi^\delta y)^i,$$

where h_i are power series with coefficients in the algebraic closure of R, and all of these coefficients are divisible by the small power π^δ. Furthermore, for each positive integer n, we can write

$$y^n = \sum_{i=0}^{d-1} \varphi_{n,i}(x) y^i$$

where $\varphi_{n,i}$ is a polynomial of degree $\ll n$ ($\leq$ some constant times n, the constant depending only on f). It then follows at once that there exists ε such that if we write

$$w^n = \sum_{i=0}^{d-1} \sum_{j=0}^{\infty} c_{j,i}^{(n)} x^j y^i$$

then ord $c_{j,i}^{(n)} \geq \varepsilon(j + n)$. This proves the lemma.

We shall say that V_0 or f_0 is **special** if $f_0'(y_0)^{-1} \in k[x, y_0]$. It then follows that

$$f'(y)^{-1} \in \mathscr{A}.$$

123

Indeed, let $z \in \mathscr{A}$ be such that $z \equiv f_0'(y_0)^{-1} \bmod \pi$. Then

$$zf'(y) = 1 + w,$$

where $w \in \mathscr{A}$ and $w \equiv 0 \bmod \pi$. Applying Lemma 1.3 proves our assertion.

Lemma 1.4. *Assume that f_0 is special. Then $\mathscr{A}$ is integrally closed in $R\langle\!\langle x \rangle\!\rangle_p[y]$, and also in the quotient field K of $\mathscr{A}$.*

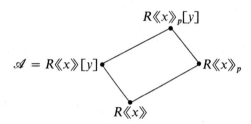

Proof. The powers y^j $(j = 0, \ldots, d - 1)$ form a basis of K over the quotient field of $R\langle\!\langle x \rangle\!\rangle$. The dual basis with respect to the trace is contained in $f'(y)^{-1}\mathscr{A}$, and hence in $\mathscr{A}$. If an element $z \in R\langle\!\langle x \rangle\!\rangle_p[y]$ is integral over $\mathscr{A}$, and we write

$$z = \sum_{i=0}^{d-1} g_i(x)y^i \quad \text{with } g_i(x) \in R\langle\!\langle x \rangle\!\rangle_p,$$

then the coefficients $g_i(x)$ can be expressed as traces,

$$g_i(x) = \mathrm{Tr}(zy_i'),$$

where $\{y_1', \ldots, y_d'\}$ is the dual basis. Since each zy_i' and its conjugates are integral over $R\langle\!\langle x \rangle\!\rangle$, so is the trace, which lies in $R\langle\!\langle x \rangle\!\rangle$ by Lemma 1.2. Hence $z \in R\langle\!\langle x \rangle\!\rangle[y]$. The same argument shows that $\mathscr{A}$ is integrally closed. (All of this is standard elementary theory of fields and integral closure, as in algebraic number theory.)

We shall need to lift roots of polynomials, as expressed in the next lemma. We let Q be the quotient field of $R\langle\!\langle x \rangle\!\rangle$.

Lemma 1.5. *Let V be a lifting of V_0 as above, and assume that V_0 is special. Let*

$$\mathscr{A} = R\langle\!\langle x \rangle\!\rangle[y] \to k[x, y_0]$$

be reduction mod π. *To each root $\bar{z}$ of f_0 in $k[x, y_0]$ there exists a unique root z of f in $\mathcal{A}$ such that*

$$z \bmod \pi = \bar{z}.$$

If in addition $k(x, y_0)$ is Galois over $k(x)$ with group G_0, then $Q(y)$ is Galois over Q. Say G is its Galois group. Then the map

$$\sigma \mapsto \sigma \bmod \pi$$

gives an isomorphism of G with G_0.

Proof. Let $z_1 \in \mathcal{A}$ reduce to $\bar{z}$ mod π. Such z_1 exists because reduction mod π gives a surjective homomorphism of $\mathcal{A}$ onto $k[x, y_0]$. We can find $w \in \mathcal{A}$ such that

$$wf'(z_1) \equiv 1 \bmod \pi.$$

We then define the usual sequence

$$z_{n+1} = z_n - wf(z_n).$$

We obtain

$$f(z_{n+1}) = f(z_n) - wf(z_n)f'(z_n) \bmod f(z_n)^2.$$

It follows by induction that

$$f(z_n) \equiv 0 \bmod \pi^n \quad \text{and} \quad z_n \equiv z_1 \bmod \pi.$$

The limit z of the sequence $\{z_n\}$ then a priori lies in the ring

$$R\langle\!\langle x \rangle\!\rangle_p[y],$$

and is a root of f in that ring. By Lemma 1.4, we conclude that the root actually lies in $R\langle\!\langle x \rangle\!\rangle[y] = \mathcal{A}$, as desired.

Under the additional assumption of Galois extensions as stated, this proves that $\mathcal{A}$ is Galois over $R\langle\!\langle x \rangle\!\rangle$. Since the prime π remains a prime ideal in $\mathcal{A}$, by standard decomposition group arguments we see that G is the decomposition group of the prime, and we get the isomorphism of G with G_0 by reduction mod π. Cf. Proposition 14, Chapter I, §5 of my *Algebraic Number Theory*.

The last lemma is only a special case of a more general situation having to do with the possibility of lifting morphisms. What we use in connection with the Frobenius morphism is summarized in the next lemma.

Lemma 1.6. *Let W_0, V_0 be two special affine curves, defined over k by polynomials g_0, f_0 respectively. Let W, V be liftings, defined by g, f. Let*

$$\varphi_0 : \mathscr{A}_0(V_0) \to \mathscr{A}_0(W_0)$$

be a homomorphism of the affine rings, such that

$$\varphi_0(x) \in xk[x].$$

Then there exists a lifting homomorphism

$$\varphi : \mathscr{A}(V) \to \mathscr{A}(W)$$

of φ_0. If $\varphi(x) \in R\langle\!\langle x \rangle\!\rangle$ is a lifting of $\varphi_0(x)$, then there is a unique choice of $\varphi(y)$ lifting $\varphi_0(y)$ and making φ a p-adically continuous homomorphism.

Proof. Suppose $\varphi(x)$ lifts $\varphi_0(x)$. Then we can define a unique p-adically continuous homomorphism $R\langle\!\langle x \rangle\!\rangle \to R\langle\!\langle x \rangle\!\rangle$ by

$$\sum a_n x^n \mapsto \sum a_n \varphi(x)^n.$$

Let φf be the polynomial obtained by applying φ to the coefficients of f, and let $(\varphi f)_0$ be its reduction mod π. Then $(\varphi f)_0$ has a root $\bar{z}$ in $\mathscr{A}_0(W_0)$, namely $\bar{z} = \varphi_0(y_0)$. Let $z_1 \in \mathscr{A}(W)$ reduce to $\bar{z}$ mod π. As in the proof of Lemma 1.5 we then find w in $\mathscr{A}$ such that

$$w(\varphi f)'(z_1) \equiv 1 \,(\mathrm{mod}\ \pi),$$

and the same argument as in Lemma 1.5 then shows that z_1 can be uniquely refined to a root of φf in $\mathscr{A}(W)$. This concludes the proof.

§2. The Artin–Schreier Equation

The example we shall consider is given by the equation

$$Y^p - Y = x^N$$

where N is a positive integer prime to p. If k is a field of characteristic p, then $Y^p - Y - x^N$ is irreducible. (For instance, a root is ramified of order p over $x = \infty$.) Furthermore, putting $f_0(Y) = Y^p - Y - x^N$ (as a polynomial over k), we have

$$f_0'(Y) = pY^{p-1} - 1 = -1.$$

Viewing now

$$f(Y) = Y^p - Y - x^N$$

as a polynomial in characteristic 0, i.e. with coefficients in R, then for the root y of $f(Y)$ we have

$$f'(y) = py^{p-1} - 1.$$

We see that f_0 (or the affine curve V_0) is special as we defined it in the last section, and that it satisfies all the hypotheses of the lemmas, which are therefore applicable. We call V_0 the **Artin–Schreier curve**.

The equation $Y^p - Y - x^N = 0$ has a formal solution both in characteristic p and characteristic 0. Indeed, it has the approximate solution $Y = 0 \bmod x$, and since $f'(0) = -1$, the Newton sequence of approximate solutions converges in the formal power series to a unique solution $h(x)$ such that $h(0) = 0$. In fact,

$$h(x) = -x^N + \cdots .$$

In characteristic p, it has the group of automorphisms $G_0(p)$ isomorphic to $\mathbf{Z}/p\mathbf{Z}$, sending

$$y_0 \mapsto y_0 + \alpha, \quad \text{for } \alpha \in \mathbf{Z}(p).$$

By Lemma 1.5, there is a unique automorphism σ_α of $\mathscr{A}$ leaving $R\langle\!\langle x \rangle\!\rangle$ fixed such that

$$\sigma_\alpha(y) \equiv y + \alpha \,(\mathrm{mod}\ \pi).$$

We let $G(p)$ be the lifting of $G_0(p)$ as a group of automorphisms of $\mathscr{A}$, as in Lemma 1.6. The map $\alpha \mapsto \sigma_\alpha$ is an isomorphism.

On the other hand, we have a lattice of rings:

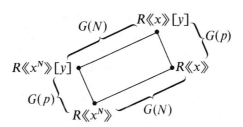

From now on, assume that R contains the N-th roots of unity. Then $R\langle\!\langle x \rangle\!\rangle$ is Galois over $R\langle\!\langle x^N \rangle\!\rangle$ with group $G(N)$ isomorphic to a cyclic group of order N, sending

$$x \mapsto \zeta x \quad \text{for } \zeta \in \mathbf{\mu}_N.$$

The two extensions corresponding to the two bottom sides of the parallelogram are linearly disjoint (being of relatively prime degree), and so opposite sides of the parallelogram have isomorphic Galois groups. In particular, the ring

$$\mathcal{A} = R\langle\!\langle x \rangle\!\rangle [y]$$

admits the group

$$G = G(p) \times G(N)$$

as a group of automorphisms, whose fixed ring is $R\langle\!\langle x^N \rangle\!\rangle$. The fixed rings of $G(p)$ and $G(N)$ are $R\langle\!\langle x \rangle\!\rangle$ and $R\langle\!\langle x^N \rangle\!\rangle [y]$ respectively.

The character group $\hat{G}$ consists of character (ψ, χ), where ψ is a character on $\mathbf{Z}(p)$ and χ is a character which may be identified with a character on roots of unity. We write

$$\chi = \chi_j$$

if $\chi(\zeta) = \zeta^j$, with $0 \le j \le N - 1$.

If (ψ, χ) is a character of G, and M is a G-module, we let $M(\psi, \chi)$ be the eigenspace corresponding to this character.

Lemma 2.1. *We have $\mathcal{A}(1, \chi_j) = x^j R\langle\!\langle x^N \rangle\!\rangle$, for $0 \le j \le N - 1$.*

Proof. Obvious.

We now let $\pi^{p-1} = -p$ and

$$R = \mathbf{Z}_p[\pi, \boldsymbol{\mu}_N].$$

In the previous chapter, §3, we had associated with each π such that $\pi^{p-1} = -p$ a character ψ_π of $\mathbf{Z}(p)$, and the Dwork power series

$$E_\pi(X) = \exp(\pi X - \pi X^p).$$

It follows from Lemma 2.2 of the preceding chapter that

$$E_\pi(y) \in R\langle\!\langle y \rangle\!\rangle \subset \mathcal{A}.$$

Lemma 2.2. *The element $E_\pi(y)$ is an eigenvector of $G(p)$ with eigencharacter ψ_π. In other words, for $\alpha \in \mathbf{Z}(p)$,*

$$\sigma_\alpha(E_\pi(y)) = \psi_\pi(\alpha)E_\pi(y).$$

Furthermore, $E_\pi(y)$ is a unit in $R\langle\!\langle y \rangle\!\rangle$.

Proof. For p odd, the inverse of $E_\pi(y)$ is $E_\pi(-y)$, which is thus also in $R\langle\!\langle y\rangle\!\rangle$. If $p = 2$, one has to apply Lemma 1.3. Next we verify that $\sigma_\alpha E_\pi(y)$ and $E_\pi(y)$ differ by a p-th root of unity. We have:

$$
\begin{aligned}
(\sigma_\alpha E_\pi(y))^p &= \sigma_\alpha(E_\pi(y)^p) \\
&= \sigma_\alpha \exp(p\pi y - p\pi y^p) \\
&= \sigma_\alpha \exp(-\pi p x^N) \\
&= \exp(-\pi p \sigma_\alpha x^N) = \exp(-p\pi x^N) = E_\pi(y)^p.
\end{aligned}
$$

Hence

$$
\frac{\sigma_\alpha E_\pi(y)}{E_\pi(y)} \in \mu_p.
$$

The Galois theory of Lemma 1.5 applied to

$$
R\langle\!\langle x^N\rangle\!\rangle[y] = R\langle\!\langle y\rangle\!\rangle
$$

shows that σ_α is an automorphism of $R\langle\!\langle y\rangle\!\rangle$. Hence $\sigma_\alpha E_\pi(y)$ is a power series in y, and in fact,

$$
\begin{aligned}
\sigma_\alpha E_\pi(y) = E_\pi(\sigma_\alpha y) &\equiv 1 + \pi\sigma_\alpha y \,(\mathrm{mod}\ \pi^2) \\
&\equiv 1 + \pi y + \pi\alpha \,(\mathrm{mod}\ \pi^2).
\end{aligned}
$$

On the other hand,

$$
E_\pi(y) \equiv 1 + \pi y \,(\mathrm{mod}\ \pi^2).
$$

Taking the quotient shows that

$$
\frac{\sigma_\alpha E_\pi(y)}{E_\pi(y)} \equiv 1 + \alpha\pi \,(\mathrm{mod}\ \pi^2).
$$

This proves that the quotient on the left-hand side is equal to $\psi_\pi(\alpha)$, as desired.

The units $\sigma_\alpha E_\pi(y)$ will allow us to determine an eigenspace decomposition for $\mathscr{A}$, except that we need p in the denominator of the orthogonal idempotents of the group ring $G(p)$. Hence we let K be the quotient field of R, so that

$$
K = R[1/p].
$$

Then we shall abbreviate by $\mathscr{A}_K$ the ring

$$
\mathscr{A}_K = \mathscr{A}[1/p] \approx \mathscr{A} \otimes K.
$$

For each $\alpha \in \mathbf{Z}(p)$ we suppose given a unit $E_\alpha \in \mathscr{A}$ such that E_α is an eigenvector for $G(p)$ with eigenvalue $\psi_\pi(\alpha)$. The family of units $\sigma_\alpha E_\pi(y)$ forms one example, but there is another, equally natural, namely the family of units

$$E_\pi(y)^i \quad \text{with } i = 0, \ldots, p - 1.$$

Since $E_\pi(y)$ has eigenvalue $\psi_\pi(1)$, it follows that $E_\pi(y)^i$ has eigenvalue $\psi_\pi(1)^i$.

Theorem 2.3. *We have the eigenspace decomposition*

$$\mathscr{A}_K = \bigoplus_{\alpha \in \mathbf{Z}(p)} E_\alpha \cdot R\langle\!\langle x \rangle\!\rangle_K$$

or also

$$\mathscr{A}_K = R\langle\!\langle x \rangle\!\rangle_K \oplus \bigoplus_{i=1}^{p-1} E_\pi(y)^i R\langle\!\langle x \rangle\!\rangle_K.$$

Proof. Note that $\mathscr{A}_K$ is free of dimension p over $R\langle\!\langle x \rangle\!\rangle_K$. Each eigenvector provides for a one-dimensional subspace, and hence their sum (necessarily direct) is the whole space $\mathscr{A}_K$.

The fact that $R\langle\!\langle x \rangle\!\rangle$ is the fixed subring of $G(p)$ implies that the eigenspace for the trivial character is precisely

$$R\langle\!\langle x \rangle\!\rangle_K.$$

Recall that not only $G(p)$ acts on $\mathscr{A}$ but also $G(N)$, and so the group

$$G = G(p) \times G(N).$$

A character of $G(N)$ can be viewed as χ_j with $0 \leq j \leq N - 1$, and χ_j is non-trivial if and only if $1 \leq j \leq N - 1$.

Theorem 2.4. *Let $0 \leq i \leq p - 1$ and $0 \leq j \leq N - 1$. Then*

$$\mathscr{A}_K(\psi_\pi^i, \chi_j) = x^j E_\pi(y)^i R\langle\!\langle x^N \rangle\!\rangle_K.$$

Proof. Obvious from Lemma 2.1 and Lemma 2.2, and the fact that $E_\pi(y)$ is fixed under $G(N)$.

Remark. The use of the Dwork power series $E_\pi(X)$ to obtain eigenspace decompositions on spaces associated with the Artin–Schreier curve is due to Monsky, cf. [Dw 5], last section of the paper.

§3. Washnitzer–Monsky Cohomology

Cohomology of differential forms with growth conditions on the coefficients was considered long ago by Dwork [Dw 1], [Dw 2]. The particular cohomology considered here is due to Washnitzer–Monsky [Wa–M], [M 1], [M 2], following the work of Dwork.

The K-algebra $\mathscr{A}_K$ is finite separable over $R\langle\!\langle x \rangle\!\rangle_K$. We have the ordinary differentiation d/dx on the power series in x. This differentiation extends in a unique way to $\mathscr{A}$, because from the relation $f(x, y) = 0$ with coefficients in R, we get

$$D_1 f(x, y)\, dx + D_2 f(x, y)\, dy = 0,$$

so

$$dy = -\frac{D_1 f(x, y)}{D_2 f(x, y)}\, dx.$$

Here, if $z \in \mathscr{A}_K$ then dz denotes the functional on derivations arising from the pairing

$$(z, D) \mapsto Dz.$$

Since we are dealing with a special curve, we know that $D_2 f(x, y)$ is invertible in $\mathscr{A}$. Consequently the above formula expressing dy in terms of dx is valid over $\mathscr{A}$. (For an elementary discussion of the foundations of the theory of derivations, cf. my *Algebra*, Chapter X, §7.)

We let

$$\Omega(\mathscr{A}) = \mathscr{A}\, dx \quad \text{and} \quad \Omega(\mathscr{A}_K) = \Omega_K(\mathscr{A}) = \mathscr{A}_K\, dx$$

be the spaces of 1-forms. We let the **Washnitzer–Monsky cohomology group** be

$$H_K^1(\mathscr{A}) = H^1(\mathscr{A}_K) = \Omega_K/d\mathscr{A}_K.$$

Note that the differential d acts like 0 on K, so

$$d(cz) = c\, dz \quad \text{for } z \in \mathscr{A},\, c \in K.$$

We are interested in the eigenspace decomposition of $H_K^1(\mathscr{A})$ with respect to $G = G(p) \times G(N)$. We note that the differential

$$\frac{dx}{x}$$

is invariant under G. The theorems of the last section give an eigenspace decomposition into three pieces:

$$\mathscr{A}_K = R\langle\!\langle x \rangle\!\rangle_K \oplus \bigoplus_{i=1}^{p-1} E_\pi(y)^i R\langle\!\langle x^N \rangle\!\rangle_K \oplus \bigoplus_{i=1}^{p-1}\bigoplus_{j=1}^{N-1} E_\pi(y)^i x^j R\langle\!\langle x^N \rangle\!\rangle_K.$$

These three pieces correspond to:

  (i) trivial action by $G(p)$;
  (ii) trivial action by $G(N)$;
  (iii) direct sum of $\mathscr{A}_K(\psi_\pi^i, \chi_j)$ with both ψ_π^i and χ_j non-trivial.

In abbreviated notation, this direct sum can be written

$$\mathscr{A}_K = \mathscr{A}_K^{G(p)} \oplus \mathscr{A}_{K,0}^{G(N)} \oplus \bigoplus_{\substack{\psi \neq 1 \\ \chi \neq 1}} \mathscr{A}_K(\psi, \chi),$$

where $\mathscr{A}_{K,0}^{G(N)}$ is the second piece in the above direct sum. Note that

$$\mathscr{A}_K^{G(p)} \cap \mathscr{A}_K^{G(N)} = R\langle\!\langle x^N \rangle\!\rangle_K,$$

whence the need for a subdivision of $\mathscr{A}_K^{G(N)}$ into two smaller subspaces to be able to write the direct sum as above.

We now want a similar decomposition for Ω_K.

Lemma 3.1. *We have*:

  (i) $\Omega_K^{G(p)} = R\langle\!\langle x \rangle\!\rangle_K\, dx = \mathscr{A}_K^{G(p)}\, dx$;
  (ii) $\Omega_K^{G(N)} = \mathscr{A}_K^{G(N)} \dfrac{dx}{x} \cap \Omega_K$;
  (iii) *for non-trivial* ψ, χ *we have*

$$\Omega_K(\psi, \chi) = \mathscr{A}_K(\psi, \chi)\frac{dx}{x}.$$

Proof. Write a differential form as

$$\sum_{i=0}^{p-1} g_i(x)y^i\, dx.$$

Invariance under $G(p)$ implies that $g_i = 0$ if $i \geq 1$, and conversely, so (i) is clear. If ζ is a primitive N-th root of unity, then $d(\zeta x) = \zeta dx$. Hence invariance under $G(N)$ implies that

$$\zeta g_i(\zeta x) = g_i(x) \quad \text{for } i = 0, \ldots, p - 1,$$

or equivalently

$$g_i(\zeta x) = \zeta^{-1} g_i(x).$$

If $N = 1$ then assertion (ii) is clear. If $N > 1$, then this last relation implies that $g_i(x)$ has no constant term, and that in the power series expansion, all terms are zero except those involving x^n with $n \equiv -1 \pmod{N}$. Then (ii) is also clear in this case. For (iii), we write an element of $\Omega_K(\psi, \chi)$ in the form

$$g(x, y) \frac{dx}{x}$$

with $g(x, y) \in \mathscr{A}_K$. Since dx/x is invariant under G. it follows that

$$g(x, y) \in \mathscr{A}_K(\psi, \chi),$$

and $g(x, y)$ is divisible by x because $j \geq 1$ if $\chi = \chi_j$, so the lemma is proved.

Lemma 3.2. *The differential operator d maps:*

$$d: \mathscr{A}_K^{G(p)} \to \Omega_K^{G(p)}$$

$$d: \mathscr{A}_K^{G(N)} \to \Omega_K^{G(N)}$$

$$d: \mathscr{A}_K(\psi, \chi) \to \mathscr{A}_K(\psi, \chi) \frac{dx}{x} = \Omega_K(\psi, \chi) \quad \text{for } \psi \neq 1, \chi \neq 1.$$

Proof. The first inclusion is clear. For the other two, we have

$$\frac{dE_\pi(y)}{E_\pi(y)} = d(\pi y - \pi y^p) = d(-\pi x^N) = -\pi N x^{N-1} \, dx.$$

Since $\varphi \mapsto d\varphi/\varphi$ is homomorphic, we get

(1) $$\frac{dE_\pi(y)^i}{E_\pi(y)^i} = -i\pi N x^N \frac{dx}{x}.$$

Using the rule for the derivative of a product, we then also easily find for any $\varphi(x) \in R\langle\!\langle x \rangle\!\rangle$:

(2) $$d(E_\pi(y)^i x^j \varphi(x^N)) = E_\pi(y)^i x^j (D_{j,i}\varphi)(x^N) N \frac{dx}{x}$$

where $D_{j,i}$ is the differential operator on a power series $\varphi(x)$ given by

$$D_{j,i} = x \frac{d}{dx} - i\pi x + \frac{j}{N}.$$

We shall work with $i = 1$, and we thus let

$$D_j = x \frac{d}{dx} - \pi x + \frac{j}{N}.$$

This proves the lemma, and in addition gives explicit formulas for the differentials, summarized by the following commutative diagram.

$$
\begin{array}{ccc}
R\langle\!\langle x\rangle\!\rangle_K & \longrightarrow & x^j E_\pi(y)R\langle\!\langle x^N\rangle\!\rangle_K \\
{\scriptstyle ND_j}\Big\downarrow & & \Big\downarrow{\scriptstyle d} \\
R\langle\!\langle x\rangle\!\rangle_K & \longrightarrow & x^j E_\pi(y)R\langle\!\langle x^N\rangle\!\rangle_K \dfrac{dx}{x}
\end{array}
$$

The horizontal map on top sends

$$\varphi(x) \mapsto x^j E_\pi(y)\varphi(x^N),$$

and similarly on the bottom, with dx/x on the right-hand side.

Recall from Theorem 2.4 that for $1 \le j \le N - 1$, we have

$$\mathscr{A}_K(\psi_\pi, \chi_j) = x^j E_\pi(y)R\langle\!\langle x^N\rangle\!\rangle_K.$$

Theorem 3.3.

(i) *For $\psi \ne 1$ and $\chi \ne 1$ we have*

$$H_K^1(\mathscr{A})(\psi, \chi) = \mathscr{A}_K(\psi, \chi)\frac{dx}{x}\Big/ d(\mathscr{A}_K(\psi, \chi)).$$

(ii) *For $1 \le j \le N - 1$, we have an isomorphism*

$$R\langle\!\langle x\rangle\!\rangle_K/D_j R\langle\!\langle x\rangle\!\rangle_K \xrightarrow{\;\approx\;} H^1(\psi_\pi, \chi_j)$$

given by

$$\varphi(x) \mapsto x^j E_\pi(y)\varphi(x^N)\frac{dx}{x}.$$

Proof. The first assertion is clear from Lemma 3.2. The second comes from the above diagram and Lemma 3.2.

The affine ring $\mathbf{Z}[x, y]$ has a natural embedding in the power series ring,

$$\mathbf{Z}[x, y] \to \mathbf{Z}[[x]],$$

mapping y on a uniquely determined power series $y = h(x)$, such that $h(0) = 0$. We have already mentioned this formal solution, and the fact that

$$h(x) = -x^N + \cdots.$$

Since $f(Y)$ is irreducible over $R\langle\!\langle x\rangle\!\rangle$ and its quotient field (for instance by the Galois theory), the root $h(x)$ of $f(Y)$ in $R[[x]]$ gives rise to the embedding

$$\mathscr{A} = R\langle\!\langle x\rangle\!\rangle[y] \to R[[x]],$$

sending y on $h(x)$. This gives rise to a natural homomorphism

$$H^1(\mathscr{A}_K) \longrightarrow H^1(R[[x]]_K)$$

$$\approx \Bigg\downarrow \qquad\qquad \Bigg\downarrow \approx$$

$$\mathscr{A}_K \, dx/d\mathscr{A}_K \qquad R[[x]]_K \, dx/dR[[x]]_K.$$

Theorem 3.4. *We have an isomorphism*

$$H^1(\mathscr{A}_K)(\psi, \chi_j) \xrightarrow{\approx} H_{j,\pi},$$

where $H_{j,\pi}$ is the representation space of the last chapter.

Proof. Clear from Theorem 3.3.

§4. The Frobenius Endomorphism

If φ is a lifting of a morphism φ_0 of special varieties, then we let $\varphi^* = H_K^1(\varphi)$ be the induced homomorphism on the cohomology groups. It can be shown that φ^* is independent of the lifting, but we shall not need this here.

We shall deal especially with automorphisms of the Artin–Schreier curve

$$\sigma = (\sigma_\alpha, \sigma_\zeta) \quad \text{with } \alpha \in \mathbf{Z}(p) \text{ and } \zeta \in \boldsymbol{\mu}_N.$$

Such $\sigma \in G = G(p) \times G(N)$ is an automorphism of $\mathscr{A} = R\langle\!\langle x\rangle\!\rangle[y]$. We have

$$\sigma^* = (\sigma_\alpha^*, \sigma_\zeta^*).$$

We also have the Frobenius endomorphism

$$F_0 : \mathcal{A}_0(V_0) \to \mathcal{A}_0(V_0)$$

such that $F_0(x, y_0) = (x^p, y_0^p)$ in characteristic p. By Lemma 1.6 we know that it can be lifted uniquely to an endomorphism

$$F : \mathcal{A} \to \mathcal{A} \quad \text{such that } F(x) = x^p,$$

extending by K-linearity to an endomorphism of $\mathcal{A}_K$.

In characteristic p, that is on V_0, we obviously have

$$(1) \qquad\qquad F \circ (\sigma_\alpha, \sigma_\zeta) = (\sigma_\alpha, \sigma_\zeta^p) \circ F.$$

Indeed, in characteristic p, $(\sigma_\alpha, \sigma_\zeta)(x, y_0) = (\zeta x, y_0 + \alpha)$, so

$$\begin{aligned}
F_0 \circ (\sigma_\alpha, \sigma_\zeta)(x, y_0) &= (\zeta^p x^p, y_0^p + \alpha^p) \\
&= (\zeta^p x^p, y_0^p + \alpha) \\
&= (\sigma_\alpha, \sigma_\zeta^p)(x^p, y^p) \\
&= (\sigma_\alpha, \sigma_\zeta^p) \circ F_0(x, y_0).
\end{aligned}$$

The commutation rule in characteristic zero follows by the uniqueness of liftings of homomorphisms when the x-value is prescribed.

From (1) we get on the cohomology

$$(1^*) \qquad\qquad (\sigma_\alpha, \sigma_\zeta)^* \circ F^* = F^* \circ (\sigma_\alpha, \sigma_\zeta^p)^*.$$

In particular, on an eigenspace we find that for $\psi \neq 1, \chi \neq 1$,

$$F^* : H_K^1(\psi, \chi) \to H_K^1(\psi, \chi^p).$$

Now we wish to see what happens to the Frobenius endomorphism under the embedding of $\mathcal{A}$ in $R[[x]]$. Mutatis mutandis, we know that $h(x^p)$ is the unique solution in power series with zero constant term of the equation

$$T^p - T = x^{pN}.$$

Theorem 3.4 now shows that the representation of F^* on the distinguished elements

$$x^j E_\pi(y) \frac{dx}{x}$$

corresponds to the representation on the elements

$$x^j \exp(-\pi x^N) \frac{dx}{x}$$

arising from the last chapter. Thus we find:

Theorem 4.1. *The eigenvalue of F_q^* on $H^1(\mathscr{A}_k)(\psi_\pi, \chi_j)$ is the same as the eigenvalue of Φ_q^* on $H_{j,\pi}$.*

17 Gauss Sums as Distributions

The Stickelberger theorem giving the factorization of Gauss sums, the Gross–Koblitz formula, and the Davenport–Hasse distribution relations will be combined to interpret Gauss sums as universal odd distributions (Yamamoto's theorem).

On the other hand, Diamond [Di 1] and Morita [Mo] gave a value of $L'_p(0, \chi)$ in terms of the p-adic gamma function. Ferrero–Greenberg gave a variation of Diamond's formula, and used it to show that $L'_p(0, \chi) \neq 0$ under the appropriate conditions. The value $L'_p(0, \chi)$ is essentially the generator for the χ-eigenspace of the Stickelberger distribution, but the proof of the non-vanishing requires the analogue of Baker's theorem (Brumer in the p-adic case), as in the proof of the non-vanishing of the p-adic regulator of cyclotomic fields. This is combined with the linear algebra of distribution relations, especially in the composite case. Cf. Kubert–Lang [KL 5].

The formula for $L'_p(0, \chi)$ will be derived by an elegant method of Washington [Wa 3], who gives the p-adic analogue of the partial zeta functions. Over the complex numbers, the coefficients in the expansion at $s = 1$ are themselves interesting functions (of gamma type), which appear thus in a natural way as homomorphic images of the partial zeta distributions. The same thing happens in the p-adic case.

§1. The Universal Distribution

We assume that the reader is acquainted with Chapter 2, §8, §9, §10. In particular, suppose that $N > 1$ is an integer. Let $(\mathbf{Q}/\mathbf{Z})_N = (1/N)\mathbf{Z}/\mathbf{Z}$, and let

$$g : (\mathbf{Q}/\mathbf{Z})_N \to A$$

be a function into some abelian group A. Such a function is called an **ordinary distribution**—**distribution** for short—if it satisfies the condition

$$\sum_{i=0}^{M-1} g\left(x + \frac{i}{M}\right) = g(Mx)$$

for every divisor M of N, and all $x \in (\mathbf{Q}/\mathbf{Z})_N$.

We let $\mathbf{F}(N)$ be the free abelian group generated by $(\mathbf{Q}/\mathbf{Z})_N$. We let $\mathbf{DR}(N)$ be the subgroup of distribution relations, that is, the subgroup generated by the elements of the form

$$\sum_{i=0}^{M-1} \left(x + \frac{i}{M}\right) - (Mx), \quad \text{for all } M \mid N.$$

We let $\mathbf{U}(N) = \mathbf{F}(N)/\mathbf{DR}(N)$ be the factor group, which we call the **universal distribution** of level N. The natural map

$$(\mathbf{Q}/\mathbf{Z})_N \to \mathbf{U}(N)$$

is then universal for distributions into abelian groups in the obvious sense. Given a distribution g on $(\mathbf{Q}/\mathbf{Z})_N$, there exists a unique homomorphism g_* making the diagram commutative:

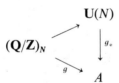

Kubert's theorem (Chapter 2, Theorem 9.2) asserts that $\mathbf{U}(N)$ is free on $\phi(N)$ generators.

Theorem 1.1. *Let g be a distribution as above. Let K be a field of characteristic 0. Assume that the distribution obtained by following g with the natural homomorphism*

$$A \to A \otimes K$$

has K-rank $\phi(N)$, in the sense that the dimension of the vector space generated by the image of $(\mathbf{Q}/\mathbf{Z})_N$ has dimension $\phi(N)$. Then g is the universal distribution.

Proof. The rank of the image is at most $\phi(N)$. If the vector space generated by the image has that rank, then the Kubert generators must remain free under g and the tensor product, so they must be linearly independent over $\mathbf{Z}$

in the abelian group generated by $g((\mathbf{Q}/\mathbf{Z})_N)$. Hence the canonical homomorphism from the universal distribution to g must be an isomorphism, as was to be shown.

Let

$$G(N) \approx \mathbf{Z}(N)^*$$

be a group isomorphic to $\mathbf{Z}(N)^*$, the isomorphism being denoted by

$$\sigma_c \leftrightarrow c.$$

We think of $G(N)$ as the Galois group of $\mathbf{Q}(\mu_N)$ over $\mathbf{Q}$. Let h be an ordinary distribution, and let as before the **Stickelberger distribution associated with h** be defined by

$$\mathrm{St}_h(x) = \mathrm{St}(x) = \sum_{c \in \mathbf{Z}(N)^*} h(xc)\sigma_c^{-1}.$$

If χ is a character of $G(N)$ with conductor m, we define

$$S(\chi, h) = S_m(\chi_m, h_m) = \sum_{c \in \mathbf{Z}(m)^*} \chi(c)h(c/m).$$

In fact, Theorem 1.1 can be made more precise, and again in Chapter 2, §8 we proved the following facts. Let $M \,|\, N$, and let χ be a character of $G(N)$. Let

$$e_\chi = \frac{1}{|G(N)|} \sum \bar{\chi}(c)\sigma_c$$

be the usual idempotent projecting on the χ-eigenspace.

ST 1. *If* cond χ *does not divide M, then*

$$\mathrm{St}_h\!\left(\frac{1}{M}\right)e_\chi = 0.$$

ST 2. *If* cond χ *divides M and has the same prime factors as M, then*

$$\mathrm{St}_h\!\left(\frac{1}{M}\right)e_\chi = \frac{|G(N)|}{|G(M)|} S(\bar{\chi}, h)e_\chi.$$

ST 3. *If* cond χ *divides M, and we let $m = $* cond χ, *then*

$$\mathrm{St}_h\!\left(\frac{1}{M}\right)e_\chi = \frac{|G(N)|}{|G(M)|} \prod_{\substack{p|M \\ p\nmid m}} (1 - \bar{\chi}(p))S(\bar{\chi}, h)e_\chi.$$

An arbitrary value $St_h(a/M)$ for a prime to M comes from the formula

$$St_h\left(\frac{a}{M}\right) = \sigma_a\, St_h\left(\frac{1}{M}\right),$$

so such values do not contain essentially more information on the image of the Stickelberger distribution than the normalized values $St_h(1/M)$. We suppose that h takes its values in an algebraically closed field K of characteristic 0, which in the applications is $\mathbf{C}$ or $\mathbf{C}_p$ for some prime p. It would suffice to suppose that h takes values in a field containing enough roots of unity, but we can always extend scalars and still be able to use Theorem 1.1. The element $S(\chi, h)$ is then an element of K.

Theorem 1.2. *Let h be a distribution with values in K. Let V be the vector space generated by the image of the Stickelberger distribution. Then $e_\chi V$ is generated by the single element $S(\bar\chi, h)$, and in particular has dimension 0 or 1 according as that element is 0 or $\neq 0$.*

Proof. The proof of Theorem 8.2 in Chapter 2 in fact proves the statement as given here, although we stated previously only the corresponding dimension property.

Corollary. *If $S(\bar\chi, h) \neq 0$ for all χ, then St_h is universal.*

Proof. This follows just like Theorem 1.1, but we don't even need to tensor with K since the values of the distribution h are already in K, and the values of the Stickelberger distribution are already in a vector space over K.

A distribution h is called **odd** or **even** according as

$$h(-x) = -h(x) \quad \text{or} \quad h(-x) = h(x).$$

From now on we restrict ourselves to distributions whose values are in abelian groups without 2-torsion. This condition will not be repeated. Such a group may then be embedded in a group where multiplication by 2 is invertible. When that is the case, any distribution is uniquely expressible as an even distribution plus an odd distribution, in the usual manner.

Theorem 1.1 then remains valid for odd (respectively even) universal distributions, except that the rank is then $\phi(N)/2$ for $N \geq 3$, which we assume.

Likewise, in Theorem 1.2, if h is, say, an odd distribution, then St_h is also odd, and is universal odd if $S(\bar\chi, h) \neq 0$ for all odd characters χ.

Example. Let

$$\begin{aligned} h(x) &= \mathbf{B}_1(\langle x\rangle) \quad \text{if } x \neq 0 \\ h(0) &= 0. \end{aligned}$$

We call h the **first Bernoulli distribution**. As we saw in Chapter 2, Theorem 8.3, its associated Stickelberger distribution is the universal odd distribution. This comes back to the fact that

$$B_{1,\chi} = S(\chi, h) \neq 0$$

for odd characters χ.

§2. The Gauss Sums as Universal Distributions

In Chapter 2, §10 we already gave the Davenport–Hasse distribution relation. Another proof of this relation follows from the Gross-Koblitz formula, and is left as an exercise for the reader. Here we are concerned with showing to what extent the Gauss sums give the universal odd distribution.

We let $h(x)$ be the first Bernoulli distribution as mentioned at the end of the last section. We pick $N = q - 1$ where $q = p^r$ for some *odd prime p*, and we have by definition

$$St(x) = \sum_c h(xc)\sigma_c^{-1}.$$

The sum is taken for $c \in \mathbf{Z}(q - 1)^*$. Write

$$x = \frac{a}{q - 1}.$$

Define

$$g(x) = \tau(\omega_q^{-a})/\tau(\omega_q^{(q-1)/2}).$$

Stickelberger's theorem gives the ideal factorization of $g(x)$, namely by Theorem 2.2 of Chapter 1, we know that

$$(g(x)) = \mathfrak{p}^{St(x)}.$$

It is convenient typographically and otherwise to write this formally additively, and thus we call

$$St(x) \cdot \mathfrak{p}$$

the **associated divisor** of $g(x)$. We view at first

$$g : (\mathbf{Q}/\mathbf{Z})_{q-1} \to \mathbf{C}_p^*$$

as a map into the multiplicative group of p-adic complex numbers. The distribution relation is then satisfied only with fudge factors. To get rid of them, we let

$P = \{\mu, p^{1/\infty}\}$ be the group generated by all roots of
unity and fractional powers of p.

We call P the **pure group** (with respect to the prime p). We may then compose g with the canonical homomorphism $C_p^* \to C_p^*/P$ to obtain a map

$$g_P : (\mathbf{Q}/\mathbf{Z})_{q-1} \to C_p^*/P$$

into the factor group, which is a distribution by the Davenport–Hasse relation. We call g_P the **Davenport–Hasse** or **Gauss sum distribution**. Note that C_p^*/P is uniquely divisible by 2. Recall that the p-adic logarithm has kernel equal precisely to the pure group P. Consequently we have a natural isomorphism of distributions

GSD 1.
$$\boxed{g_P \approx \log_p g.}$$

The map

$$St(x) \mapsto St(x) \cdot \mathfrak{p}$$

is a homomorphism of the Stickelberger distribution (which we know is universal odd). The decomposition group D_p of $\mathfrak{p}$ consists of the powers of p mod N. If we let $G(N) \approx \mathbf{Z}(N)^*$ under the notation

$$a \mapsto \sigma_a,$$

then the values of the homomorphic image above can be viewed as lying in the group ring

$$\mathbf{Q}[G(N)/D_p],$$

and

$$St_h(x) \bmod D_p = \sum_{c \in \mathbf{Z}(N)^*/D_p} \left(\sum_{i=0}^{r-1} h(p^i x c) \right) \sigma_c^{-1}.$$

Lemma 2.1. *Let* $\{m(x)\}$ *be a family of integers and let*

$$\alpha = \prod_x g(x)^{m(x)}.$$

Then α *is pure if and only if* div $\alpha = 0$, *and in that case* α *is a root of unity.*

143

Proof. The absolute value of $g(x)$ in the complex numbers is 1. Hence $|\alpha| = 1$. If α is pure, this implies that α is a root of unity. Conversely, assume that div $\alpha = 0$, so α is a unit. The conjugates of $g(x)$ also have absolute value 1 (themselves being of the same form as $g(x)$), and it is standard that a unit all of whose conjugates have absolute value 1 must be a root of unity. This proves the lemma.

From the lemma, it follows that we have an isomorphism

GSD 2.
$$\boxed{g_P \approx \operatorname{St}_h \bmod D_p .}$$

Note. Factoring out by D_p corresponds to the obvious classical fact that the Gauss sums satisfy the relation

$$\tau(\chi) = \tau(\chi^p),$$

cf. Chapter 1, **GS 4**. In the present notation, this is written

$$g(x) = g(px).$$

Let χ be an odd character of conductor d prime to p. Since the Gauss sum distribution is also odd, it follows that the function

$$a \longmapsto \chi(a) \log_p g\left(\frac{a}{d}\right)$$

is even, for $a \in \mathbf{Z}(d)^*$. Hence the function is defined on $\mathbf{Z}(d)^*/\pm 1$.

Theorem 2.2 (Ferrero–Greenberg). *Let χ be an odd character of conductor d. Assume that $\chi(p) = 1$. Then*

$$\sum_{a \in \mathbf{Z}(d)^*} \chi(a) \log_p g\left(\frac{a}{d}\right) \neq 0.$$

The proof will be based on the following lemma.

Lemma 2.3. *There exists a family of integers $\{m(d')\}$, for divisors d' of d, $d' \neq d$, having the following property. Let*

$$\xi = \operatorname{St}_h\left(\frac{1}{d}\right) + \sum_{d'} m(d') \operatorname{St}_h\left(\frac{1}{d'}\right).$$

Let R be a set of representatives for cosets of the group generated by D_p and ± 1 in $\mathbf{Z}(d)^*$. Then the elements

$$\sigma_a \xi, \qquad a \in R$$

are linearly independent over $\mathbf{Q}$ as elements of $\mathbf{Q}[G(N)/D_p]$.

Proof. As in Theorem 1.2 we look at the ψ-eigenspace for odd characters ψ such that $\psi(p) = 1$ and cond ψ divides d. It suffices to prove that we can choose the family $\{m(d')\}$ such that

$$\xi e_\psi \neq 0 \quad \text{for all } \psi.$$

First we know from **ST 2** that if cond $\psi = d$ then

$$\mathrm{St}_h\!\left(\frac{1}{d}\right) e_\psi \neq 0$$

because $B_{1,\psi} \neq 0$. Let $d_1 > d_2 > \cdots$ be the other divisors of d, unequal to d. Pick a sequence of integers $m_1 < m_2 < \cdots$ which is rapidly increasing. Then for any ψ we have

$$\mathrm{St}_h\!\left(\frac{1}{d}\right) e_\psi + \sum_i m_i \, \mathrm{St}_h\!\left(\frac{1}{d_i}\right) e_\psi \neq 0.$$

Indeed, suppose d_s is the conductor of ψ. By **ST 1** we conclude that the i-th term in the sum is 0 if $i > s$. By **ST 2** and the fact that $B_{1,\psi} \neq 0$ we know that

$$\mathrm{St}_h\!\left(\frac{1}{d_s}\right) e_\psi \neq 0.$$

If the sequence is selected increasing sufficiently fast, then this s-term dominates in the sum, which is therefore also not equal to 0, thus proving the lemma.

We return to Theorem 2.2. Let

$$\alpha = g\!\left(\frac{1}{d}\right) \prod_{d' \neq d} g\!\left(\frac{1}{d'}\right)^{m(d')},$$

with the family $\{m(d')\}$ chosen as in the lemma. By **GSD 1, GSD 2** we conclude that the elements

$$\log_p \sigma_a \alpha \quad \text{with } a \in R$$

are linearly independent over the rational numbers. By Baker's theorem (in the p-adic case, Brumer) it follows that they are linearly independent over the algebraic numbers. Therefore

$$0 \neq \sum_{a \in \mathbf{Z}(d)^*/\{D_p, \pm 1\}} \chi(a) \log_p(\sigma_a \alpha) = \sum_a \chi(a) \log_p g\left(\frac{a}{d}\right),$$

because

$$\sum_a \chi(a) \log_p g\left(\frac{a}{d'}\right) = 0$$

for each $d' \neq d$ since the conductor of χ is d. This concludes the proof of Theorem 2.2.

§3. The L-function at $s = 0$

Let d be a positive integer prime to p (where p is an *odd prime*). Let χ be a primitive even Dirichlet character with conductor d or dp. We take the values of χ to be in $\mathbf{C}_p$. We let $\omega = \omega_p$ be the Teichmuller character, and we define

$$\chi_n = \chi\omega^{-n}.$$

We know that

$$L_p(1 - n, \chi) = -(1 - \chi_n(p)p^{n-1})\frac{1}{n}B_{n, \chi_n}.$$

For $n = 1$, we obtain

$$L_p(0, \chi) = -(1 - \chi_1(p))B_{1, \chi_1}.$$

Since χ_1 is odd, we know that $B_{1, \chi_1} \neq 0$. Hence:

$$L_p(0, \chi) = 0 \quad \text{if and only if} \quad \chi_1(p) = 1.$$

The next formula is a slight variation of a formula of Diamond [Di], but expressed in terms of the p-adic gamma function itself by Ferrero–Greenberg [Fe–Gr].

Theorem 3.1. *Let χ be an even character such that χ_1 has conductor d. Then*

$$L'_p(0, \chi) = \sum_{c=1}^d \chi_1(c) \log_p \Gamma_p\left(\frac{c}{d}\right) + (1 - \chi_1(p))B_{1, \chi_1} \log_p(d).$$

We shall prove this formula in §4. Here we derive consequences. Indeed, the next theorem amounts to a p-adic analogue of Stark's conjectures in a special case. Considerably more insight in this direction was provided by Gross [Gr].

As in the preceding section, let $x = a/(q - 1)$ and let

$$\varphi(x) = \tau(\omega_q^{-a}), \qquad g(x) = \varphi(x)\varphi\left(\frac{q-1}{2}\right)^{-1}.$$

The formula of Theorem 3.1 involves the log of an analytic expression (the gamma function). We shall transform it so that it involves the log of an algebraic expression (a Gauss sum).

Theorem 3.2. *Let χ be an even character such that χ_1 has conductor d and such that $\chi_1(p) = 1$. Let D_p be the subgroup of $\mathbf{Z}(d)^*$ generated by the powers of p. Then*

$$L_p'(0, \chi) = \sum_{c \in \mathbf{Z}(d)^*/D_p} \chi_1(c) \log_p g\left(\frac{c}{d}\right) \neq 0.$$

Proof. By assumption the formula of Theorem 3.1 simplifies to

$$L_p'(0, \chi) = \sum_{c=1}^{d} \chi_1(c) \log_p \Gamma_p\left(\frac{c}{d}\right).$$

Let $d | (q - 1)$, $q = p^r$ where r is the period of p mod d. Let $G = \mathbf{Z}(q - 1)^*$. Then χ_1 is defined modulo D_p by assumption. Since

$$\Gamma_p(z)\Gamma_p(1 - z) = \pm 1,$$

it follows at once from the Gross–Koblitz formula that

$$\log_p \varphi(x) = \sum_{i=1}^{r} \log_p \Gamma_p(\langle p^i x \rangle).$$

In particular,

$$\log_p \varphi(px) = \log_p \varphi(x).$$

The sum over $c = 1, \ldots, d$ in the formula for $L_p'(0, \chi)$ is then written in the form

$$\sum_{c \in \mathbf{Z}(d)^*/D_p} \chi_1(c) \sum_{i=0}^{r-1} \log_p \Gamma_p\left(\left\langle \frac{p^i c}{d} \right\rangle\right) = \sum_{c \in \mathbf{Z}(d)^*/D_p} \chi_1(c) \log_p g\left(\frac{c}{d}\right)$$

where

$$g(x) = \varphi(x)\varphi\left(\frac{q-1}{2}\right)^{-1}$$

is the odd distribution of the preceding section. That $L'_p(0, \chi) \neq 0$ is then the main result of §2.

§4. The p-adic Partial Zeta Function

In the complex case, the Hankel transform gives an analytic continuation of the partial zeta function to the whole plane. In the p-adic case, we shall give an analogue of this partial zeta function due to Washington [Wa 3], who also pointed out to me that his formula for the p-adic L-function immediately gives the value of the derivative at $s = 0$ in terms of the gamma-type functions.

Let $x \in \mathbf{C}_p^*$ be such that $x^{-1} \equiv 0 \bmod \mathfrak{m}_p$. Let $s \in \mathbf{Z}_p$. We define the **Hurwitz–Washington** function

$$H(s, x) = \sum_{j=0}^{\infty} \binom{1-s}{j} x^{-j} B_j,$$

where B_j is the j-th Bernoulli number. Since the Bernoulli numbers have bounded denominator at p (Kummer and von Staudt congruences), it is easily shown that $H(s, x)$ is holomorphic for $s \in \mathbf{Z}_p$. For an integer $k \geq 1$, we find

H 1. $\qquad\qquad H(1 - k, x) = x^{-k}\mathbf{B}_k(x),$

where $\mathbf{B}_k$ is the k-th Bernoulli polynomial. This is immediate from the value of this polynomial,

$$\mathbf{B}_k(X) = \sum_{j=0}^{k} \binom{k}{j} X^{k-j} B_j$$

which comes directly from the definitions in terms of the generating power series, product of e^{tX} and $t/(e^t - 1)$.

Let N be a positive integer divisible by p. For $a \in \mathfrak{o}_p^*$ we therefore obtain

$$H\left(1 - k, \frac{a}{N}\right) = N^k a^{-k}\mathbf{B}_k\left(\frac{a}{N}\right).$$

H 2. *If f is a function on $\mathbf{Z}(N)$, then*

$$\frac{1}{N} \sum_{\substack{a=1 \\ p \nmid a}}^{N-1} f(a)a^k H\left(1 - k, \frac{a}{N}\right) = B_{k,f} - p^{k-1}B_{k,f \circ p}$$

where $(f \circ p)(x) = f(px)$ and

$$B_{k,f} = N^{k-1} \sum_{0}^{N-1} f(a)\mathbf{B}_k\left(\left\langle \frac{a}{N} \right\rangle\right).$$

This is immediate by taking the sum over all $a = 0, \ldots, N - 1$ and subtracting the sum over $a = py$, with $0 \le y \le (N/p) - 1$. It is convenient to use the following notation. Put

$$M_p f(x) = f(px), \qquad f_k = f\omega^{-k}.$$

Then **H 2** can be written in the form

H 3. $\dfrac{1}{N} \displaystyle\sum_{\substack{a=1 \\ p \nmid a}}^{N-1} f(a)\langle a\rangle_p^k \dfrac{1}{k} H\left(1 - k, \dfrac{a}{N}\right) = \displaystyle\int (1 - p^{k-1}M_p)f_k \, dE_k.$

The formula expresses the Bernoulli distribution in terms of H, giving the possibility of analytic continuation.

If χ is a Dirichlet character whose conductor divides N, then the preceding formula reads

H 4. $\dfrac{1}{N} \displaystyle\sum_{\substack{a=1 \\ p \nmid a}}^{N-1} \chi(a)\langle a\rangle_p^k H\left(1 - k, \dfrac{a}{N}\right) = (1 - \chi_k(p)p^{k-1})B_{k,\chi_k}.$

We now define the **Hurwitz–Washington function** in three variables,

$$H(s; a, N) = -\frac{1}{1 - s} \langle a\rangle_p^{1-s} \frac{1}{N} H\left(s, \frac{a}{N}\right),$$

for $a \in \mathbf{Z}_p^*$, $s \in \mathbf{Z}_p$, and N equal to a positive integer divisible by p. Then $H(s; a, N)$ is again holomorphic in s except at $s = 1$. It is the p-adic partial zeta function, cf. [Wa 3].

One could take the relation of the next theorem as the definition of the p-adic L-function, and thus make the present chapter independent of Chapter 12.

Theorem 4.1. *Let χ be a Dirichlet character, and let N be any multiple of the conductor of χ such that N is divisible by p. Then*

$$L_p(s, \chi) = \sum_{\substack{a=1 \\ p \nmid a}}^{N-1} \chi(a) H(s; a, N).$$

Proof. The left-hand side and the right-hand side have the same values at the negative integers, which are dense in $\mathbf{Z}_p$, and they are both holomorphic, hence they coincide.

We are here concerned with finding $L_p'(0, \chi)$. That we are dealing with a character χ is basically irrelevant, and so for any function on $\mathbf{Z}(N)$ we now define

$$L_p(s, f) = \sum_{\substack{a=1 \\ p \nmid a}}^{N-1} f(a) H(s; a, N).$$

In finding the expansion at $s = 0$, we shall meet the **Diamond function** defined by the formula

$$G_p(x) = (x - \tfrac{1}{2}) \log_p(x) - x + \sum_{j=2}^{\infty} \frac{B_j}{j(j-1)} x^{1-j},$$

cf. [Di 1]. This formula arises from the asymptotic expansion of the classical complex log gamma function. It converges p-adically for $|x| > 1$, so $G_p(x)$ is defined in that domain. We shall analyze later the relation between the Diamond function and the gamma function.

Theorem 4.2. *Let $p \neq 2$. Let N be a positive integer divisible by p and let f be a function on $\mathbf{Z}(N)$. Then*

$$L_p'(0, f) = \sum_{\substack{a=1 \\ p \nmid a}}^{N-1} f_1(a) G_p\left(\frac{a}{N}\right) + \sum_{\substack{a=1 \\ p \nmid a}}^{N-1} f_1(a) B_1\left(\frac{a}{N}\right) \log_p(N).$$

If $f = \chi$ is a Dirichlet character, then

$$L_p'(0, \chi) = \sum_{\substack{a=1 \\ p \nmid a}}^{N-1} \chi_1(a) G_p\left(\frac{a}{N}\right) + (1 - \chi_1(p)) B_{1, \chi_1} \log_p(N).$$

Proof. The desired result is an immediate consequence of the next lemma.

Lemma 4.3. *Let* $|a| > 1/p$ *and let* N *be a positive integer divisible by* p. *Then the coefficient of* s *in* $H(s; a, N)$ *is equal to*

$$\omega(a)^{-1} G_p\left(\frac{a}{N}\right) + \omega(a)^{-1} \mathbf{B}_1\left(\frac{a}{N}\right) \log_p(N).$$

Proof. We have the expansions:

$$\frac{1}{1-s} = 1 + s + \cdots$$

$$\langle a \rangle^{1-s} = \langle a \rangle (1 - s \log_p \langle a \rangle + \cdots)$$

$$\text{If } j \geq 2, \quad \binom{1-s}{j} = \frac{(-1)^{j-1}}{j(j-1)^s} + \cdots.$$

Using the fact that $B_j = 0$ for j odd, $j > 1$, we find that the coefficient of s in $H(s; a, N)$ is

$$-\frac{\langle a \rangle}{N}\left[1 - \log_p(a) + \frac{N}{2a} \log_p(a) - \sum_{j=2}^{\infty} \left(\frac{N}{a}\right)^j \frac{B_j}{j(j-1)} \right]$$

$$= \omega(a)^{-1}\left[-\frac{a}{N} + \left(\frac{a}{N} - \frac{1}{2}\right) \log_p\left(\frac{a}{N}\right) + \sum_{j=2}^{\infty} \left(\frac{a}{N}\right)^{1-j} \frac{B_j}{j(j-1)} \right.$$

$$\left. + \left(\frac{a}{N} - \frac{1}{2}\right) \log_p(N) \right]$$

$$= \omega(a)^{-1} G_p\left(\frac{a}{N}\right) + \omega(a)^{-1}\left(\frac{a}{N} - \frac{1}{2}\right) \log_p(N).$$

This proves the lemma.

Remark. In [Di 2], Diamond discusses the regularization of his function, giving rise to certain measures which are then related to the Bernoulli measures. See also Koblitz [Ko 1]. In the classical case, gamma-type functions appear as coefficients of partial zeta functions (Hurwitz functions), and we meet a similar phenomenon here.

Next, we derive some functional equations. First, for the Washington function, we get for p odd:

H 5. $$H(s; a, N) = H(s; N - a, N).$$

Proof. It suffices to prove the formula when $s = 1 - k$, $k \geq 2$ such that $k \equiv 0 \mod p - 1$, because such integers are dense in $\mathbf{Z}_p$. But then the formula is immediate from the fact that k is even (p is assumed odd), and the property

$$\mathbf{B}_k(1 - X) = (-1)^k \mathbf{B}_k(X),$$

which follows directly from the generating function for Bernoulli polynomials.

Washington has also pointed out that one can give elegant proofs for the following properties of the Diamond function by using the formalism of the H-function. We assume p odd.

$\mathbf{G}_p$ 1. $\qquad\qquad\qquad G_p(1 - x) + G_p(x) = 0.$

Proof. In Lemma 4.3 we found the coefficient of s in $H(s; a, N)$. Using **H 5**, and replacing a by $N - a$ in this coefficient, we now see that

$$G_p\!\left(\frac{a}{N}\right) + G_p\!\left(1 - \frac{a}{N}\right) = 0.$$

This is true for any positive integer N divisible by p, and any p-unit a, thus proving the formula.

$\mathbf{G}_p$ 2. $\qquad\qquad\qquad G_p(-x) + G_p(x) = -\log_p(x).$

Proof. Immediate from the power series expansion.

$\mathbf{G}_p$ 3. $\qquad\qquad\qquad G_p(1 + x) - G_p(x) = \log_p(x).$

Proof. Immediate from the preceding two properties.

Theorem 4.4. *Extend $G_p(x)$ to $\mathbf{Q}_p$ by putting $G_p(x) = 0$ if $x \in \mathbf{Z}_p$. Then for all $x \in \mathbf{Z}_p$ we have*

$$\sum_{b=0}^{p-1} G_p\!\left(\frac{x + b}{p}\right) = \log_p \Gamma_p(x).$$

Proof. Both sides are continuous and satisfy the functional equation

$$f(x + 1) = f(x) + \delta(x)\log_p x,$$

where $\delta(x) = 0$ if $x \equiv 0 \bmod p$, and $\delta(x) = 1$ otherwise. This is true for $\log_p \Gamma_p(x)$ directly from the definition of Γ_p, and is true for the other side by $\mathbf{G}_p$ 3. Hence the two functions differ by a constant. Putting $x = 0$ gives 0 on the right-hand side. By G_p 1 we conclude that the left-hand side is also equal to 0. This proves the theorem.

Theorem 4.5. *Let* χ *be a Dirichlet character such that the conductor d of* χ_1 *is not divisible by p. Then*

$$L_p'(0, \chi) = \sum_{c=1}^{d-1} \chi_1(c) \log_p \Gamma_p\left(\frac{c}{d}\right) + (1 - \chi_1(p))B_{1, \chi_1} \log_p(d).$$

Proof. Let $N = pd$. In Theorem 4.2, write

$$a = c + bd$$

with $1 \leq c \leq d - 1$ and $0 \leq b \leq p - 1$. Then $\chi_1(a) = \chi_1(c)$. If a is divisible by p then $a/N \in \mathbf{Z}_p$, so that $G_p(a/N) = 0$ in Theorem 4.4. The desired formula is then a direct consequence of Theorem 4.2 combined with the relation of Theorem 4.4, and the fact that $\log_p(p) = 0$.

The formula of Theorem 4.2 is due to Diamond. The argument used to derive the variation in Theorem 4.5 is in Ferrero–Greenberg [Fe–Gr].

For the record, we state the distribution relation for the Diamond function.

G_p **4.** *For any positive integer m and* $|x| > 1$, *we have*

$$\sum_{a=0}^{m-1} G_p\left(\frac{x + a}{m}\right) = G_p(x) - (x - \tfrac{1}{2}) \log_p(m).$$

In particular, if $m = p^r$ is a power of p, then:

G_p **5.** $$\sum_{a=0}^{p^r-1} G_p\left(\frac{x + a}{p^r}\right) = G_p(x).$$

This is a special case of G_p 4 because $\log_p(p) = 0$. The proof of G_p 4 can easily be given following a similar argument to that of Theorem 4.4, using the analyticity of $G_p(x)$ for $|x| > 1$. Our intent was to deal mostly with the cyclotomic applications, and we don't go into a systematic treatment of these *p*-adic functions.

Bibliography

[Am]   Y. AMICE, Les nombres *p*-adiques, Presse Universitaire de France, 1975.

[A–F]   Y. AMICE and J. FRESNEL, Fonctions zeta *p*-adiques des corps de nombres abéliens reels, *Acta Arith.* **XX** (1972) pp. 353–384

[A–H]   E. ARTIN and H. HASSE, Die beiden Erganzungssatze zum Reziprozitats gesetz der *l*ⁿ-ten Potenzreste im Korper der *l*ⁿ-ten Einheitswurzeln, *Abh. Math. Sem.* Hamburg 6 (1928) pp. 146–162

[Ba]   D. BARSKY, Transformation de Cauchy *p*-adique et algèbre d'Iwasawa, *Math. Ann.* **232** (1978) pp. 255–266

[Ba]   H. BASS, Generators and relations for cyclotomic units, *Nagoya Math. J.* **27** (1966) pp. 401–407.

[Bo]   M. BOYARSKY, *p*-adic gamma functions and Dwork cohomology, *Trans. AMS*, to appear.

[Br]   A. BRUMER, On the units of algebraic number fields, *Mathematika*, **14** (1967) pp. 121–124

[Ca]   L. CARLITZ, A generalization of Maillet's determinant and a bound for the first factor of the class number, *Proc. AMS* **12** (1961) pp. 256–261

[Ca–O] L. CARLITZ and F. R. OLSON, Maillet's determinant, *Proc. AMS* **6** (1955) pp. 265–269

[Co 1] J. COATES, On K_2 and some classical conjectures in algebraic number theory, *Ann. of Math.* **95** (1972) pp. 99–116

[Co 2] J. COATES, *K*-theory and Iwasawa's analogue of the Jacobian, Algebraic *K*-theory II, Springer *Lecture Notes* **342** (1973) pp. 502–520

[Co 3] J. COATES, *p*-adic *L*-functions and Iwasawa's theory, Durham conference on algebraic number theory and class field theory, 1976

[Co 4] J. COATES, Fonctions zeta partielles d'un corps de nombres totalement réel, Seminaire Delange–Pisot–Poitou, 1974–75

[C–L]   J. COATES and S. LICHTENBAUM, On *l*-adic zeta functions, *Ann. of Math.* **98** (1973) pp. 498–550

155

[C–S 1]   J. COATES and W. SINNOTT, On p-adic L-functions over real quadratic fields, *Invent. Math.* **25** (1974) pp. 253–279

[C–S 2]   J. COATES and W. SINNOTT, An analogue of Stickelberger's theorem for higher K-groups, *Invent. Math.* **24** (1974) pp. 149–161

[C–S 3]   J. COATES and W. SINNOTT, Integrality properties of the values of partial zeta functions, *Proc. London Math. Soc.* **1977** pp. 365–384

[C–W 1]   J. COATES and A. WILES, Explicit Reciprocity laws, Proceedings, Conference at Caen, *Soc. Math. France Astrisque* **41–42** (1977) pp. 7–17

[C–W 2]   J. COATES and A. WILES, On the conjecture of Birch and Swinnerton-Dyer, *Invent. Math.* **39** (1977) pp. 223–251

[C–W 3]   J. COATES and A. WILES, Kummer's criterion for Hurwitz numbers, Kyoto Conference on Algebra Number Theory, 1977

[C–W 4]   J. COATES and A. WILES, On the conjecture of Birch–Swinnerton-Dyer II, to appear

[Col]   R. COLEMAN, Some modules attached to Lubin–Tate groups, to appear

[D–H]   H. DAVENPORT and H. HASSE, Die Nullstellen der Kongruenz-zetafunktionen in gewissen zyklischen Fällen, *J. reine angew. Math.* **172** (1935) pp. 151–182

[Di 1]   J. DIAMOND, The p-adic log gamma function and p-adic Euler constant, *Trans. Am. Math. Soc.* **233** (1977) pp. 321–337

[Di 2]   J. DIAMOND, The p-adic gamma measures, *Proc. AMS* **75** (1979) p. 211–217.

[Di 3]   J. DIAMOND, On the values of p-adic L-functions at positive integers, *Acta Aithm XXXV* (1979) pp. 223–237.

[Dw 1]   B. DWORK, On the zeta function of a hypersurface, *Publ. Math. IHES* **12** (1962) pp. 5–68

[Dw 2]   B. DWORK, A deformation theory for the zeta function of a hypersurface, *Proc. Int. Cong. Math.* Stockholm, 1962

[Dw 3]   B. DWORK, On the zeta function of a hypersurface II, *Ann. of Math.* **80** (1964) pp. 227–299

[Dw 4]   B. DWORK, p-adic cycles, *Publ. Math. IHES* **37** (1969) pp. 28–115

[Dw 5]   B. DWORK, Bessel functions as p-adic functions of the argument, *Duke Math. J.* **41** No. 4 (1974) pp. 711–738

[Dw–Ro]   B. DWORK and P. ROBBA, On ordinary linear p-adic differential equations, *Trans. AMS* Vol. **231** (1977) pp. 1–46

[En 1]   V. ENNOLA, On relations between cyclotomic units, *J. Number Theory* **4** (1972) pp. 236–247

[En 2]   V. ENNOLA, Some particular relations between cyclotomic units, *Ann. Univ. Turkuensis* **147** (1971)

[Fe 1]   B. FERRERO, Iwasawa invariants of abelian number fields, *Math. Ann.* **234** (1978) pp. 9–24

[Fe 2]   B. FERRERO, An explicit bound for Iwasawa's λ-invariant, *Acta Arithm.* **XXXIII** (1977) pp. 405–408

[Fe–Gr]   B. FERRERO and R. GREENBERG, On the behavior of p-adic L-functions at $s = 0$, to appear

[Fe–W]   B. FERRERO and L. WASHINGTON, The Iwasawa invariant μ_p vanishes for abelian number fields, *Ann. Math.* **109** (1979) pp. 377–395

[Fre]   J. FRESNEL, Nombres de Bernoulli et fonctions L p-adiques, *Ann. Inst. Fourier* **17** (1967) pp. 281-333

[Fro]   A. FROHLICH, Formal groups, Springer *Lectures Notes in Mathematics* **74**, 1968

[Gi 1] R. GILLARD, Unités cyclotomiques, unités semi-locales et Z_l-extensions, *Ann. Inst. Fourier*, to appear

[Gi 2] R. GILLARD, Extensions abéliennes et répartition modulo 1, *Journées arithmétiques de Marseille*, Asterisque, 1979.

[Gr 1] R. GREENBERG, A generalization of Kummer's criterion, *Invent. Math.* **21** (1973) pp. 247-254

[Gr 2] R. GREENBERG, On a certain l-adic representation, *Invent. Math.* **21** (1973) pp. 117-124

[Gr 3] R. GREENBERG, The Iwasawa invariants of Γ-extensions of a fixed number field, *Amer. J. Math.* **95** (1973) pp. 204-214

[Gr 4] R. GREENBERG, On the Iwasawa invariants of totally real number fields, *Amer. J. Math.* **93** (1976) pp. 263-284

[Gr 5] R. GREENBERG, On p-adic L-functions and cyclotomic fields, *Nagoya Math. J.* **56** (1974) pp. 61-77

[Gr 6] R. GREENBERG, On p-adic L-functions and cyclotomic fields II, to appear

[Gr 7] R. GREENBERG, A note on K_2 and the theory of Z_p-extensions, to appear

[Gr]   B. GROSS, On the behavior of p-adic L-functions at s = 0, to appear.

[Gr-Ko] B. GROSS and N. KOBLITZ, Gauss sums and the p-adic gamma function, *Ann. of Math.* **109** (1979) pp. 569-581.

[Ha 1] H. HASSE, Uber die Klassenzahl abelschen Zahlkorper, Akademie Verlag, Berlin, 1952

[Ha 2] H. HASSE, Bericht...., Teil II, Reziprozitätsgesetz, Reprinted, Physica Verlag, Würzburg, Wien, 1965

[Ha 3] H. HASSE, Theorie der relativ zyklischen algebraischen Funktionenkörper insbesonderen bei endlichen Konstantenkörper, *J. reine angew. Math.* **172** (1935) pp. 37-54

[Ho]   T. HONDA, On the formal structure of the Jacobian variety of the Fermat curve over a p-adic integer ring, Symposia Mathematica Vol. **XI**, Academic Press, NY, 1973, pp. 271-284

[Hu-V] L. K. HUA and H. S. VANDIVER, *Proc. Nat. Acad. Sci. USA* **34** (1948) pp. 258-263

[Iw 1] K. IWASAWA, On Γ-extensions of algebraic number fields, *Bull. Amer. Math. Soc.* **65** (1959) pp. 183-226

[Iw 2] K. IWASAWA, A note on the group of units of an algebraic number field, *J. Math. pures et app.* **35** (1956) pp. 189-192

[Iw 3] K. IWASAWA, Sheaves for algebraic number fields, *Ann. of Math.* **69** (1959) pp. 408-413

[Iw 4] K. IWASAWA, On some properties of Γ-finite modules, *Ann. of Math.* **70** (1959) pp. 291-312

[Iw 5] K. IWASAWA, On the theory of cyclotomic fields, *Ann. of Math.* **70** (1959) pp. 530-561

[Iw 6] K. IWASAWA, On some invariants of cyclotomic fields, *Amer. J. Math.* **80** (1958) pp. 773-783

[Iw 7]   K. IWASAWA, A class number formula for cyclotomic fields, *Ann. of Math.* **76** (1962) pp. 171–179

[Iw 8]   K. IWASAWA, On some modules in the theory of cyclotomic fields, *J. Math. Soc. Japan* Vol. 16, No. 1 (1964) pp. 42–82

[Iw 9]   K. IWASAWA, Some results in the theory of cyclotomic fields, *Symposia in Pure Math.* Vol. VIII, AMS 1965, pp. 66–69

[Iw 10]   K. IWASAWA, On explicit formulas for the norm residue symbol, *J. Math. Soc. Japan* Vol. 20, Nos. 1–2 (1968) pp. 151–165

[Iw 11]   K. IWASAWA, On p-adic L-functions, *Ann. of Math.* **89** (1969) pp. 198–205

[Iw 12]   K. IWASAWA, On Z_l-extensions of algebraic number fields, *Ann. of Math* **98** (1973) pp. 246–326

[Iw 13]   K. IWASAWA, A note of cyclotomic fields, *Invent. Math.* **36** (1976) pp. 115–123

[Iw 14]   K. IWASAWA, Lectures on p-adic L-functions, *Ann. of Math. Studies* No. 74

[Iw 15]   K. IWASAWA, On the μ-invariants of Z_l-extensions, Conference on number theory, algebraic geometry, and commutative algebra in honor of Y. Akizuki, Tokyo (1973) pp. 1–11

[Iw 16]   K. IWASAWA, Some remarks on Hecke characters, *Algebraic Number Theory*, Kyoto International Symposium, 1976, Japan Society for Promotion of Science, Tokyo 1977, pp. 99–108

[Iw 17]   K. IWASAWA, A note of Jacobi sums, *Symposia Mathematica XV*, (1975) pp. 447–459

[Iw 18]   K. IWASAWA, Analogies between number fields and function fields, see *Math. Reviews* Vol. **41**, No. 172

[Jo]   W. JOHNSON, Irregular primes and cyclotomic invariants, *Math. Comp.* **29** (1975) pp. 113–120

[Ka 1]   N. KATZ, Formal groups and p-adic interpolation, to appear

[Ka 2]   N. KATZ, On the differential equations satisfied by period matrices, *Publ. Math. IHES* **35** (1968) pp. 71–106

[Ka 3]   N. KATZ, Une formule de congruence pour la fonction zeta, SGA 7 II, Expose 22 p. 401, Springer *Lecture Notes* **340**

[Ka 4]   N. KATZ, Another look at p-adic L-functions for totally real fields, to appear.

[Ko 1]   N. KOBLITZ, Interpretation of the p-adic log gamma function and Euler constants using the Bernoulli measures, *Trans. AMS* (1978) pp. 261–269.

[Ko 2]   N. KOBLITZ, A new proof of certain formulas for p-adic L-functions, *Duke J.* (1979) pp. 455–468.

[Ku 1]   D. KUBERT, A system of free generators for the universal even ordinary $Z_{(2)}$ distribution on Q^{2k}/Z^{2k}, *Math. Ann.* **224** (1976) pp. 21–31

[Ku 2]   D. KUBERT, The Universal ordinary distribution, *Bull. Soc. Math. France* **107** (1979) pp. 179–202.

[KL 1]   D. KUBERT and S. LANG, Units in the modular function field, I, Diophantine applications, *Math. Ann.* **218** (1975) pp. 67–96

[KL 2]   D. KUBERT and S. LANG, Idem II, A full set of units, pp. 175–189

[KL 3]   D. KUBERT and S. LANG, Idem III, Distribution relations, pp. 273–285

[KL 4]   D. KUBERT and S. LANG, Idem IV, The Siegel functions are generators, *Math. Ann.* **227** (1977) pp. 223–242

[KL 5]   D. KUBERT and S. LANG, Distributions on toroidal groups, *Math. Zeit.* (1976) pp. 33–51

[KL 6]   D. KUBERT and S. LANG, The *p*-primary component of the cuspidal divisor class group on the modular curve *X*(*p*), *Math. Ann.* **234** (1978) pp. 25–44

[KL 7]   D. KUBERT and S. LANG, The index of Stickelberger ideals of order 2 and cuspidal class numbers, *Math. Ann.* **237** (1978) pp. 213–232

[KL 8]   D. KUBERT and S. LANG, Stickelberger ideals, *Math. Ann.* **237** (1978) pp. 203–212

[KL 9]   D. KUBERT and S. LANG, Iwasawa theory in the modular tower, *Math. Ann.* **237** (1978) pp. 97–104.

[Ku–L]   T. KUBOTA and H. LEOPOLDT, Eine *p*-adische Theorie der Zetawerte, *J. reine angew. Math.* **214/215** (1964) pp. 328–339

[L 1]   S. LANG, *Algebraic Number Theory*, Addison-Wesley, 1970

[L 2]   S. LANG, *Elliptic functions*, Addison-Wesley, 1973

[L 3]   S. LANG, *Introduction to modular forms*, Springer-Verlag, 1976

[L 4]   S. LANG, *Elliptic curves: Diophantine analysis*, Springer-Verlag, 1978

[Leh]   D. H. LEHMER, *Applications of digital computers*, in *Automation and Pure Mathematics*, Ginn, Boston (1963) pp. 219–231

[Le 1]   H. W. LEOPOLDT, Zur Geschlechtertheorie in abelschen Zahlkörpern, *Math. Nach.* **9**, 6 (1953) pp. 351–363

[Le 2]   H. W. LEOPOLDT, Uber Einheitengruppe und Klassenzahl reeller Zahlkörper, Abh. Deutschen Akad. Wiss. Berlin, Akademie Verlag, Verlag, Berlin, 1954

[Le 3]   H. W. LEOPOLDT, Über ein Fundamentalproblem der Theorie der Einheiten algebraischer Zahlkörper, Sitzungsbericht Bayerischen Akademie Wiss. (1956), pp. 41–48

[Le 4]   H. W. LEOPOLDT, Eine Verallgemeinerung der Bernoullischen Zahlen, *Abh. Math. Sem. Hamburg* (1958) pp. 131–140

[Le 5]   H. W. LEOPOLDT, Zur Struktur der *l*-Klassengruppe galoisscher Zahlkörper, *J. reine angew. Math.* (1958) pp. 165–174

[Le 6]   H. W. LEOPOLDT, Uber Klassenzahlprimteiler reeller abelscher Zahlkörper, *Abh. Math. Sem. Hamburg* (1959) pp. 36–47

[Le 7]   H. W. LEOPOLDT, Uber die Hauptordnung der ganzen Elemente eines abelschen Zahlkörpers, *J. reine angew. Math.* **201** (1959) pp. 113–118

[Le 8]   H. W. LEOPOLDT, Uber Fermatquotienten von Kreiseinheiten und Klassenzahlformeln modulo *p*, *Rend. Circ. Mat. Palermo* (1960) pp. 1–12

[Le 9]   H. W. LEOPOLDT, Zur approximation des *p*-adischen Logarithmus, *Abh. Math. Sem. Hamburg* **25** (1961) pp. 77–81

[Le 10]   H. W. LEOPOLDT, Zur Arithmetik in abelschen Zahlkörpern, *J. reine angew. Math.* **209** (1962) pp. 54–71

[Le 11]   H. W. LEOPOLDT, Eine *p*-adische Theorie der Zetawerte II, *J. reine angew Math.* **274–275** (1975) pp. 224–239

[Li 1]   S. LICHTENBAUM, Values of zeta functions, étale cohomology, and algebraic *K*-theory, Algebraic *K*-Theory II, Springer-Verlag *Lecture Notes in Mathematics* **342** (1973) pp. 489–501

159

[Li 2]   S. LICHTENBAUM, Values of zeta and L-functions at zero, *Soc. Math. France Asterisque* 24–25 (1975)

[Li 3]   S. LICHTENBAUM, On p-adic L-functions associated to elliptic curves, to appear

[Lu]   J. LUBIN, One parameter formal Lie groups over p-adic integer rings, *Ann. of Math.* **80** (1964) pp. 464–484

[L–T]   J. LUBIN and J. TATE, Formal complex multiplication in local fields, *Ann. of Math.* **8** (1965) pp. 380–387

[Man]   J. MANIN, Cyclotomic fields and modular curves, *Russian Math. Surveys* Vol. 26, No. 6, Nov-Dec 1971, pp. 7–78

[Mas 1]   J. MASLEY, On Euclidean rings of integers in cyclotomic fields, *J. reine angew. Math.* **272** (1975) pp. 45–48

[Mas 2]   J. MASLEY, Solution of the class number two problem for cyclotomic fields, *Invent. Math.* **28** (1975) pp. 243–244

[M–M]   J. MASLEY and H. MONTGOMERY, Cyclotomic fields with unique factorization, *J. reine angew. Math.* **286** (1976) pp. 248–256

[Maz]   B. MAZUR, Analyse p-adique, Bourbaki report, 1972

[M–SwD]   B. MAZUR and H. SWINNERTON-DYER, Arithmetic of Weil curves, *Invent. Math.* **18** (1972) pp. 183–266

[Me]   T. METSANKYLA, Class numbers and μ-invariants of cyclotomic fields, *Proc. Am. Math. Soc.* **43** No. 2 (1974) pp. 299–300

[Mi]   J. MILNOR, Introduction to algebraic K-theory, *Ann. of Math. Studies* **72** (1971)

[Mi]   H. MITCHELL, The generalized Jacobi–Kummer function, *Transactions AMS* (1916) pp. 165–177

[Mo]   Y. MORITA, A p-adic analogue of the Γ-function, *J. Fac. Sci. Tokyo* Section 1A Vol. **22** (1975) pp. 255–266

[No.]   A. P. NOVIKOV, Sur le nombre de classes des extensions abeliennes d'un corps quadratique imaginaire, *Izv. Akad. Nauk SSSR* **31** (1967) pp. 717–726

[Oe]   J. OESTERLE, Bourbaki report on Ferrero–Washington, Bourbaki Seminar, February 1979

[Pol]   F. POLLACZEK, Uber die irregulären Kreiskorper der l-ten ond l²-ten Einheitswürzeln, *Math. Zeit.* **21** (1924) pp. 1–38

[Qu 1]   D. QUILLEN, Finite generation of the groups K_i of rings of algebraic integers, Algebraic K-Theory 1, Springer *Lecture Notes* **341** (1973) pp. 179–198

[Qu 2]   D. QUILLEN, Higher algebraic K-theory I, in Algebraic K-theory I, Springer *Lecture Notes in Mathematics* **341** (1973) pp. 85–147

[Ra]   K. RAMACHANDRA, On the units of cyclotomic fields, *Acta Arith.* **12** (1966) pp. 165–173

[Ri]   K. RIBET, A modular construction of unramified p-extensions of $\mathbf{Q}(\mu_p)$, *Invent. Math.* **34** (1976) pp. 151–162

[Ro]   G. ROBERT, Nombres de Hurwitz et unités élliptiques, *Ann. scient. Ec. Norm. Sup.* 4e série t. 11 (1978) pp. 297–389.

[S–T]   A. SCHOLZ and O. TAUSSKY, Die Hauptideale der kubischen Klassenkörper imaginär-quadratischer Zahlkörper, *J. reine angew. Math.* **171** (1934) pp. 19–41

[Se 1]   J.-P. SERRE, Classes des corps cyclotomiques, d'après Iwasawa, Seminaire Bourbaki, 1958

[Se 2]   J.-P. SERRE, Formes modulaires et fonctions zeta *p*-adiques, Modular functions in one variable III, Springer *Lecture Notes* **350** (1973)

[Se 3]   J.-P. SERRE, Endomorphismes complètement continus des espaces de Banach *p*-adiques, *Pub. Math.* IHES **12** (1962) pp. 69–85

[Se 4]   J.-P. SERRE, Sur le résidu de la fonction zeta *p*-adique d'un corps de nombres, *C. R. Acad. Sci. France* t. **287** (1978) pp. 183–188

[Sh]   T. SHINTANI, On evaluation of zeta functions of totally real algebraic number fields at non-positive integers, *J. Fac. Sci. Univ. Tokyo* IA Vol. 23, No. 2 (1976) pp. 393–417

[Si 1]   C. L. SIEGEL, Uber die Fourierschen Koeffizienten von Modulformen, *Göttingen Nachrichten* **3** (1970) pp. 15–56

[Si 2]   C. L. SIEGEL, Zu zwei Bemerkungen Kummers, *Nachr. Akad. Wiss. Gottingen* **6** (1964) pp. 51–57

[Sin]   W. SINNOTT, On the Stickelberger ideal and the circular units, *Annals of Mathematics* **198** (1978) pp. 107–134

[St]   H. STARK, L-functions at $s = 1$, III: Totally real fields and Hilbert's twelfth problem, *Adv. Math.* Vol. 22, No. 1 (1976) pp. 64–84

[Ta 1]   J. TATE, Letter to Iwasawa on a relation between K_2 and Galois cohomology, in Algebraic K-Theory II, Springer *Lecture Notes* **342** (1973) pp. 524–527

[Ta 2]   J. TATE, Relations between K_2 and Galois cohomology, *Invent. Math.* **36** (1976) pp. 257–274

[Ta 3]   J. TATE, Symbols in arithmetic, *Actes Congrès Intern. Math.* 1970, Tome 1, pp. 201–211

[Va 1]   H. S. VANDIVER, Fermat's last theorem and the second factor in the cyclotomic class number, *Bull. AMS* **40** (1934) pp. 118–126

[Va 2]   H. S. VANDIVER, Fermat's last theorem, *Am. Math. Monthly* **53** (1946) pp. 555–576

[Wag]   S. WAGSTAFF, The irregular primes to 125,000, *Math. Comp.* **32**, (1978) pp. 583–591

[Wa 1]   L. WASHINGTON, Class numbers of elliptic function fields and the distribution of prime numbers, *Acta Arith.* **XXVII** (1975) pp. 111–114

[Wa 2]   L. WASHINGTON, Class numbers and Z_p-extensions, *Math. Ann.* **214** (1975) pp. 177–193

[Wa 3]   L. WASHINGTON, A note on *p*-adic L-functions, *J. Number Theory* **8** Vol. 2 (1976) pp. 245–250

[Wa 4]   L. WASHINGTON, On Fermat's last theorem, *J. reine angew. Math.* **289** (1977) pp. 115–117

[Wa 5]   L. WASHINGTON, Units of irregular cyclotomic fields, *Ill. J. Math.* to appear

[Wa 6]   L. WASHINGTON, The class number of the field of 5^nth roots of unity, *Proc. AMS*, **61** 2 (1976) pp. 205–208

[Wa 7]   L. WASHINGTON, The calculation of $L_p(1, \chi)$, *J. Number Theory*, **9** (1977) pp. 175–178

Bibliography

[Wa 8]   L. WASHINGTON, Euler factors for p-adic L-functions, *Mathematika* **25** (1978) pp. 68–75

[Wa 9]   L. WASHINGTON, The non-p part of the class number in a cyclotomic $\mathbf{Z}_p$-extension, *Inv. Math.* **49** (1979), pp. 87–97

[Wa 11]   L. WASHINGTON, Kummer's calculation of $L_p (1, \chi)$, *J. reine angew. Math.* **305** (1979) pp. 1–8

[Wa 12]   L. WASHINGTON, The derivative of p-adic L-functions, to appear

[We 1]   A. WEIL, Number of solutions of equations in finite fields, *Bull. AMS* **55** (1949) pp. 497–508

[We 2]   A. WEIL, Jacobi sums as Grossencharaktere, *Trans. AMS* **73** (1952) pp. 487–495

[We 3]   A. WEIL, Sommes de Jacobi et caracteres de Hecke, *Gött. Nach.* (1974) pp. 1–14

[We 4]   A. WEIL, On some exponential sums, *Proc. Nat. Acad. Sci. USA* **34**, No. 5 (1948) pp. 204–207

[Wi]   A. WILES, Higher explicit reciprocity laws, *Ann. of Math.* 107 (1978) pp. 235–254

[Ya 1]   K. YAMAMOTO, The gap group of multiplicative relationships of Gaussian sums, *Symposia Mathematica* No. 15, (1975) pp. 427–440

[Ya 2]   K. YAMAMOTO, On a conjecture of Hasse concerning multiplicative relations of Gaussian sums, *J. Combin. Theory* **1** (1966) pp. 476–489

Index